The Patentability of Software

This book explores the question of whether software should be patented. It analyses the ways in which the courts of the US, the EU, and Australia have attempted to deal with the problems surrounding the patentability of software and describes why it is that the software patent issue should be dealt with as a patentable subject matter issue, rather than as an issue of novelty or non-obviousness.

Anton Hughes demonstrates that the current approach has failed and that a fresh approach to the software patent problem is needed. The book goes on to argue against the patentability of software based on its close relationship to mathematics. Drawing on historical and philosophical accounts of mathematics in pursuit of a better understanding of its nature and focusing the debate on the conditions necessary for mathematical advancement, the author puts forward an analytical framework centred around the concept of the useful arts. This analysis both explains mathematics', and therefore software's, non-patentability and offers a theory of patentable subject matter consistent with Australian, American, and European patent law.

Anton Hughes has a doctorate in patent law and currently practices as a barrister in Sydney, Australia.

Routledge Research in Intellectual Property

The Patentability
of Software
Software as Mathematics

Anton Hughes

LONDON AND NEW YORK

a GlassHouse book

First published 2019
by Routledge
2 Park Square, Milton Park, Abingdon, Oxon OX14 4RN

and by Routledge
605 Third Avenue, New York, NY 10017

First issued in paperback 2021

Routledge is an imprint of the Taylor & Francis Group, an informa business

Publisher's Note
The publisher has gone to great lengths to ensure the quality of this reprint but points out that some imperfections in the original copies may be apparent.

British Library Cataloguing-in-Publication Data
A catalogue record for this book is available from the British Library

Library of Congress Cataloging-in-Publication Data
Names: Hughes, Anton, (Lawyer), author.
Title: The patentability of software : software as mathematics / Anton Hughes.
Description: New York, NY : Routledge, 2019. | Series: Routledge research in intellectual property | Includes index.
Identifiers: LCCN 2018049138 (print) | LCCN 2018049460 (ebook) | ISBN 9781315283210 (ebk) | ISBN 9781138240599 (hbk)
Subjects: LCSH: Software protection—Law and legislation. | Patent laws and legislation. | Computer software—Law and legislation. | Mathematics.
Classification: LCC K1443.C6 (ebook) | LCC K1443.C6 H84 2019 (print) | DDC 346.048/6—dc23
LC record available at https://lccn.loc.gov/2018049138

Typeset in Galliard
by Apex CoVantage, LLC

ISBN 13: 978-1-03-224141-8 (pbk)
ISBN 13: 978-1-138-24059-9 (hbk)

DOI: 10.4324/9781315283210

Contents

Acknowledgements

Having returned to the topic of my PhD thesis after some five years was a daunting topic, not just because the pendulum has swung in a new(ish) direction since I wrapped this project up last time. I was lucky enough to have a front row seat for at least one of those changes, appearing for the Commissioner in *Research Affiliates*. I must thank my then tutor, Christian Dimitriadis, and my leader, David Catterns QC, for the opportunity to contribute to that case.

I could never have completed this update without support from my family – Sarah, Alice, and Niamh – who didn't seem to mind me taking extra time out of their lives to finish it off. So a special thank you to them.

I also thank Routledge for giving me the opportunity to bring my argument to a wider audience, for their understanding when it took me longer than I expected, and for nursing me through the process of getting a book published for the first time.

Nor could I have made it across the line without the support of all those who supported me in completing my PhD. Because of that, I have repeated the acknowledgements from my PhD thesis below.

Acknowledgements from PhD thesis

First and foremost, I have to thank my wife Sarah, who has physically, emotionally, and intellectually supported me through this entire process. Aside from forbidding the discussion of mathematics after 10pm, Sarah has listened attentively on myriad occasions as I voiced barely comprehensible diatribes on topics whose relevance is peculiar to me alone. Thank you also for digging deep in the final run to the finish, and proof reading with the sort of dedication I have come to expect from you.

I am also greatly indebted to my supervisor, Professor Dianne Nicol, for her selfless dedication and support. Despite resolute denials as to expertise in the areas of mathematics and software, Di has at every step of the way provided insightful and encouraging feedback which motivated me to work harder, and reach higher.

I would also like to thank the Faculty of Law for extending me a scholarship, and giving me the resources I need to get to work. I have always been afforded generous and unquestioning support by the past and present Deans, Don Chalmers and Margaret Otlowski, and the staff and postgraduates of the Faculty have always made me feel welcome.

A sincere thank you must also go to the Honourable Justice Porter, who afforded me the opportunity of a front row seat for two years to watch the law unfold. Thanks to your your guidance and support, I learned much about thinking in issues, attention to detail and the value of a blank piece of A3 paper.

Thank you to my mother Desma, who understood from personal experience that the process of writing a PhD could be difficult, and whose advice about the process of finishing helped me greatly in the final stages. Thanks for all that hard work proof reading at short notice, and the positive encouragement that went with it. Thank you to my sister Natalya for the advice about aesthetics and semiotics – we truly had the coolest conversations ever.

A special thank you to Alice Sorcha and Niamh Milly, who only arrived after this project started, but have made my days and nights delightful ever since. Thanks for your support.

And finally, to my father Bob, who inspired me to want to start a thesis even before I really knew what one was. Twenty-five years later that project is complete. Thanks for the many words of wisdom that, although said almost in passing, somehow resound at just the right moments.

Introduction

In modern life, software has become ubiquitous. The primary vehicle by which software has entered the public consciousness is the personal computer. It is not unusual for people to have a computer at work and a computer at home. But people interact with and even depend on software in increasingly varied ways. Software runs on all manner of devices including "clothes washers and dryers; toasters and microwave ovens; electronic organisers; digital televisions and digital audio recorders; home alarm systems and elderly medic alert systems; irrigation systems; pacemakers; video games; Web-surfing devices; copying machines; calculators; toothbrushes; musical greeting cards; pet identification tags; and toys".[1] The modern mobile phone is a pocket computer,[2] and its software forms a key point of differentiation for many buyers.[3] Cars are increasingly reliant on software components, embedded in small microcontrollers, to manage emissions and fuel economy; for advanced diagnostics; to simplify manufacture and design; as well as for comfort and convenience features.[4] It will not be too long before the computer takes over entirely from human drivers.[5] Complex software applications drive cutting-edge scientific research projects,

1 Martin L. Shooman, *Reliability of computer systems and networks: fault tolerance, analysis and design* (John Wiley and Sons, 2002) at 9.
2 For example, the now seemingly ancient iPhone 4 had approximately the same processing power as a Cray-2 Supercomputer, the fastest machine in the world on its release in 1985. See "Processing Power Compared," <http://pages.experts-exchange.com/processing-power-compared/> (27 February 2017); "Cray-2", *Wikipedia*, last modified 23 January 2017 <https://en.wikipedia.org/wiki/Cray-2> (27 February 2017).
3 Buyers are increasingly interested in whether their phones run on iOS, Android or Windows Phone OS. See for example Denis Gallagher, "Smartphone buying guide" *Choice Online*, 30 August 2011 <https://www.choice.com.au/electronics-and-technology-phones/mobile-phones/buying-guides/smartphones> (27 February 2017).
4 Ben Wojdyla, "How it Works: The Computer Inside Your Car" on *Popular Mechanics* <http://www.popularmechanics.com/cars/how-to/a7386/how-it-works-the-computer-inside-your-car/> (27 February 2017). A typical new car in 2013 had about 100,000,000 lines of code: Doug Newcomb, "The Next Big OS War is in Your Dashboard," *Wired Magazine*, <https://www.wired.com/2012/12/automotive-os-war/> (27 February 2017).
5 See for example, William Messner, "It's pedal to the metal for driverless cars", *The Conversation*, 30 January 2017, <https://theconversation.com/its-pedal-to-the-metal-for-driverless-cars-71936> (27 February 2017).

for example the Large Hadron Collider[6] and Human Genome Project.[7] Software systems are used to manage and control the delivery of infrastructure services such as electricity, water, telecommunications and public transport. In addition to the wide range of contexts in which software is used, the size and complexity of individual programs has grown exponentially.[8] In 1981, the MS-DOS operating system which powered most personal computers comprised around 4000 lines of code.[9] By 2002, its descendant, Windows XP, contained in the region of 40,000,000 lines of code.[10]

Correspondingly, the software industry is big business. According to the US Bureau of Labor Statistics, in 2008 there were around 1.3 million computer systems software engineers and programmers in the US alone. That number was projected to increase by 21%, to around 1.6 million, by 2018.[11] The size of the global software industry was estimated to be worth US$ 225.5 billion in 2009.[12] By 2013, Gartner estimated its value at US$ 407.3 billion.[13] It

---

6 For an overview, see Wikipedia, "Large Hadron Collider" <http://en.wikipedia.org/wiki/Large_Hadron_Collider> (4 September 2011). The Large Hadron Collider depends software to control the operation of the multitude of hardware components, as well as to collect and analyse data. For an overview of the software involved, see Juan Batiz-Benet, Xuwen Cao and Yin Yin Wu, "CERN's Large Hadron Collider" <https://sites.google.com/site/multinationalsoftwares/multinationalprojects/cern-lhc> (4 September 2011).

7 The goals of the HGP were to "identify all the approximately 20,000–25,000 genes in human DNA; determine the sequences of the 3 billion chemical base pairs that make up human DNA; store this information in databases; improve tools for data analysis; transfer related technologies to the private sector; and address the ethical, legal, and social issues (ELSI) that may arise from the project": United States Department of Energy, "About the Human Genome Project" <http://www.ornl.gov/sci/techresources/Human_Genome/about.shtml> (4 September 2011). The project was completed in 2003, although the data analysis continues. On the role of automation in the project, see Lynn Yaris, "Machines and the Human Genome Project" <www.lbl.gov/Science-Articles/Archive/human-genome-mapping-sequencing.html> (4 September 2011): "'From the beginning, laboratory automation has been recognized as an essential element of the Human Genome Project,' says Ed Theil a computer systems engineer who has been working with the Human Genome Center's instrumentation group. 'Beyond the advantages of speed and relief from tedium, automation minimizes human errors and captures data instantly as part of the process.'"

8 On the relationship between size and complexity, see Chapter 1, at 22.

9 Henry M. Walker, *The tao of computing* (Jones and Bartlett Learning, 2004) at 79.

10 Henry M. Walker, *The tao of computing* (Jones and Bartlett Learning, 2004) at 79. It is acknowledged of course that the number of lines of code in an application is a very crude measure, as it does not take account of language choice, or programming style, both of which may produce marked variations in lines of code used for similar tasks.

11 United States Bureau of Labor Statistics, "Computer Software Engineers and Computer Programmers" in *Occupational Outlook Handbook, 2010–11 Edition* <http://www.bls.gov/oco/ocos303.htm> (3 September 2011).

12 DataMonitor, "Software: Global Industry Guide 2010" 31 January 2011 <http://www.datamonitor.com/store/Product/software_global_industry_guide_2010?productid=4F026C5C-EBCC-4193-AD27-77260196E7F5> (3 September 2011).

13 "Gartner Says Worldwide Software Market Grew 4.8 Percent in 2013," *Gartner.com*, Press Release 31 March 2014 <www.gartner.com/newsroom/id/2696317> (27 February 2017).

is forecast that the value of the industry will be around US$ 899.5 billion by 2019.[14]

But the software business is not like other businesses. Much of the reason for that stems from the unusual nature of the product. Software is "infinitely malleable".[15] It is an arrangement of concepts, constructed in words. Because of this, it is not engineered like traditional physical products. It does not progress sequentially through various phases from design, to prototype, to testing, to production, with the laws of physics throwing up unexpected difficulties at each stage along the way. Unlike a car, any part of a software product can be completely changed at any point along the development pipeline, and updates continue to be released after the product is distributed. No matter how much cost is involved in producing the first copy, subsequent copies can be made for free. Software might be built for one customer or a million: the development process is the same. Software is regularly built to order, bought off the shelf, or a hybrid of the two – an off-the-shelf product selected, then customised for the specific requirements of an individual or an organisation.

The flexibility and freedom of software is both its strength and its weakness. The ability to modify a product at any point, even after it has been released on the open market, offers users the potential to adapt software to their needs, rather than adapting their needs to the software. But the temptation to change can be hard to resist. This can make project requirements a constantly moving target. As a result, software projects are notorious for blown budgets and late delivery.

The software "factory" is also unusual. Despite the size and complexity of software creations, they can be built by comparatively small teams.[16] Those teams can share an office or be scattered around the globe. Members of the team, the programmers, might see themselves as software engineers, but they may well consider themselves artists. Given the intangibility of the medium they work with, it can be hard to measure their productivity. Empirical evidence suggests a ten fold variation, even amongst people with similar experience.[17]

In light of these many unique features, it seems sensible to question whether software should be patentable. Traditionally, the patent system has been an

14 Marketline, "Software: Global Industry Guide," *ReportLinker*, May 2015 <http://www.reportlinker.com/p0191925> (27 February 2017).

15 Frederick P. Brooks Jr, "No Silver Bullet: Essence and Accidents of Software Engineering" (1987) 20(4) *Computer* 10 at 12.

16 "I've been heard word processors seriously compared with the complexity of Boeing's airplanes. But according to the product manager of Microsoft Word (version 3.1), his software development team consisted of only eight programmers.": Brad Cox, "No Silver Bullet Revisited" *American Programmer Journal*, November 1995 <http://virtualschool.edu/cox/pub/NoSilverBulletRevisted> (22 July 2011).

17 H. Sackman, W. J. Erickson and E. E. Grant, "Exploratory Experimental Studies Comparing Online and Offline Programming Performance" (1968) 11(1) Communications of the ACM 3; Bill Curtis, "Fifteen Years of Psychology in Software Engineering: Individual Differences and Cognitive Science" ICSE '84 Proceedings of the 7th international conference on Software engineering (IEEE Press, 1984).

important mechanism for benefiting society through the encouragement of innovation. Patent law is said to achieve such encouragement in three ways: the award of monopolies which encourages both the creation of new inventions and their commercialisation; through the disclosure of those technologies to the public, which encourages innovation in related areas; and through the availability for further development of disclosed inventions by others, after the expiry of the patent term.[18]

But patent law came into being early in the 15th century, and took its modern form towards the end of the end of the 19th century. Its original purpose was to incentivise the establishment of new industries at the start of the Industrial Revolution,[19] a time where inventing and commercialising new products was a slow process of trial and error. Patent law protected the inventor's good idea against the open market to allow it to be commercialised. Since that time, advances in science and technology have seen the patent system broaden to cover new varieties of less and less tangible invention, from chemical processes to methods of improving devices, methods of medical treatment, and even the process of isolating biological materials.[20]

Software is a highly intangible and complex creation, and its industry is characterised by a rapid product lifecycle and a high degree of dependence on reuse and conformity. Thus when software's patentability first fell to be considered in the 1960s, it raised a the question of whether its protection was a bridge too far for the patent regime. The early indications were that it was. In 1966 a US Presidential Commission composed of leading academics, industry representatives, and the Commissioner of Patents recommended that Congress legislatively exclude software from patent law.[21] In Europe, just such an exclusion of computer programs forms part of the European Patent Convention[22] and has been given legislative operation in the UK since 1977.[23] In Australia, early decisions of the Australian Patent Office were clearly against software patents, although the patentability of software was not judicially considered until 1991.[24]

Despite this early reticence, the patenting of software became commonplace in the US, the UK, and Australia. But this does not mean that software's patentability is a settled issue. The path towards software's patentability was a gradual one,

---

18 Kathy Bowrey, Michael Handler and Dianne Nicol, *Australian Intellectual Property: Commentary, Law and Practice* (Oxford University Press, 2010) at 378.

19 A classic example of this Industrial Era invention is Watt's steam engine, an improvement of which was the subject of *Boulton v Bull* (1795) 126 ER 651.

20 Although the isolation of gene sequences appears to be beyond protection: see *Association for Molecular Pathology v Myriad Genetics Inc* 569 U.S. ____, 133 S.Ct. 2107 (US Supreme Court, 13 June 2013); *D'Arcy v Myriad Genetics Inc & Anor* [2015] HCA 35 (Australian High Court, 7 October 2015).

21 Presidential Commission on the Patent System, *To Promote the Progress of Useful Arts in an Age of Exploding Technology* (1966).

22 *Convention on the Grant of European Patents*, opened for signature 5 October 1973, 13 ILM 268 (entered into force 7 October 1973).

23 *Patents Act 1977* (UK).

24 *IBM v Commissioner of Patents* (1991) 33 FCR 218.

which was arguably not completed in the US until the decision in *State Street*[25] in 1998. But the campaign against software patenting remained vociferous and determined. The widespread protests against the EU Software Directive in 2005, seen to allow for the uninhibited patenting of software in that jurisdiction, eventually won the day and collected nearly half a million signatures along the way.[26] Even now, outraged members of the blogosphere, including many software developers, rail against what they see as a broken patent system and the ludicrousness of software patenting. The outbreak of the smartphone software patent wars, pitting iPhones against Androids, only added to the discontent.[27]

In more recent times, the pendulum has swung sharply away from the patentability of software. In *Alice Corp.*,[28] the US Supreme Court held that claims to a computer-implemented escrow service for facilitating financial transactions was patent ineligible, because it was drawn to an abstract idea and no amount of computer-implementation could transform it into something else. Similarly, in Australia, the Full Federal Court in *Research Affiliates*[29] and *RPL Central*,[30] whilst not excluding software from the patent regime entirely, drew a distinction between abstract ideas on the one hand and patent-eligible technical innovations on the other.

So despite the passage of over 50 years since the patentability of software was first considered, the question of software's patentability remains indeterminate. Misunderstandings as to software's nature, based on intuition rather than analysis, have combined with a preparedness to expand traditional understandings of inherent patentability to accommodate it. This has led to practical problems for the software industry, where patents are used less as an aid to innovation and more as a defensive mechanism. This accommodating attitude also presents theoretical challenges to the coherence of the patent regime. On one level, software is an intellectual creation of the kind more familiar to the copyright regime. On another level, software controls the operation of a functional device – a

---

25 *State Street Bank & Trust Co v Signature Financial Group Inc* 149 F.3d 1368 (Fed Cir, 1998).
26 A repeat performance seems likely, over the proposed introduction of a unitary patent system in the EU. See the discussion below.
27 See for example Chloe Albanesius, "Infographic: Who Will Win the Patent Wars?" *PCMag.com*, 2 September 2011 <http://www.pcmag.com/article2/0,2817,2392375,00.asp> (4 September 2011); L Gordon Crovitz, "Google, Motorola and the Patent Wars" *The Wall Street Journal*, 22 August 2011 <http://online.wsj.com/article/SB1000142405311190 3639404576518493092643006.html> (4 September 2011); Holman W Jenkins, "Obama and the Smartphone Wars" *The Wall Street Journal*, 24 August 2011 <http://online.wsj.com/article/SB10001424053111903327904576526130093390612.html> (4 September 2011); Kimberlee Weatherall, "Samsung Galaxy Tab vs Apple iPad: the tablet patent wars hit Australia" *The Conversation*, 3 August 2011 <http://theconversation.edu.au/samsung-galaxy-tab-vs-apple-ipad-the-tablet-patent-wars-hit-australia-2660> (4 September 2011).
28 *Alice Corp. v CLS Bank International*, 573 U.S. __, 134 S. Ct. 2347 (US Supreme Court, 2014).
29 *Research Affiliates LLC v Commissioner of Patents* [2014] FCAFC 150; 227 FCR 378.
30 *Commissioner of Patents v RPL Central Pty Ltd* [2015] FCAFC 177; 238 FCR 27.

computer – and such a machine seems a comfortable fit within patent law's historical protection of all manner of useful devices. The problem, put simply, is that software threatens to stretch the boundaries of patentable subject matter into uncharted territory – the realm of the abstract, intellectual creation. This may be the coal-face of innovation in an Information Age, but whether patent law's Industrial Era regime provides an appropriate incentive is far from clear.

In all of this, software's relationship with mathematics has been overlooked. Software is a by-product of a quest for truth in mathematics, having developed out of symbolic logic and mathematical formalism. Symbolic logic seeks to reduce verifiability of mathematical results to the mechanical application of a sequence of rules starting from agreed axioms. Mathematics and software are also isomorphic activities, in that the activities involved in doing mathematics and the activities involved in developing software closely correspond to each other. This point is related to the last, in that both disciplines seek to reduce knowledge to a logical sequence of steps – one for the purpose of verification, the other for instructing a computer. This close relationship makes the status of mathematics in patent law an interesting avenue of inquiry. Unlike software, mathematics has a history arguably as long as humanity itself. So considering the patentability of mathematics, which seems always to have been non-patentable, offers a rich repository from which to draw in deciding this contemporary issue.[31]

1 Aims

Briefly stated, the aims of this book are as follows:

1 To argue that software ought not be patentable.
2 In making that argument, to incorporate considerations that go beyond a utilitarian, or purely economic, analysis.
3 To shed light on the nature of software, and the way in which it is constructed, in order to better inform debate, both in the context of intellectual property and in other legal contexts upon which such an understanding depends.

This book is primarily an argument against the patentability of software. It will be shown that the nature of software, and perhaps more importantly, the way it is created, makes it of a kind which cannot be reconciled with the history, purpose, and operation of the patent regime. But it is not the first argument advanced for that purpose, nor is it likely to be the last.

Therefore, it is intended that the argument be differentiated from others not by the ends, but by the means. Boyle notes that much of the critical IP

---

31 "[A] page of history is worth a volume of logic": *New York Trust Co v Eisner* 256 US 345 (1921) per Holmes J; "In this bright future, you can't forget your past.": Bob Marley, "No Woman, No Cry", *Live!* (1975).

scholarship takes place in a "relentlessly utilitarian framework".[32] As such, patents are often justified on purely economic grounds. Patent theory asserts that "where the norm of free competition would result in free riding by competitors and less reason to invest in new technologies, innovation is encouraged by providing innovators with the exclusive rights to their inventions."[33] Innovation is considered to be a positive on the basis that the public benefits from the disclosure of new and useful technologies, which fall into the public domain at the expiry of the patent grant period – the so-called social contract theory of patent law.[34] These benefits must be contrasted with the award of monopoly rights, which amount to a tax on society

> in two different ways: first, by the high price of goods, which, in the case of a free trade, they could buy much cheaper; and, secondly, by their total exclusion from a branch of business which it might be both convenient and profitable for many of them to carry on.[35]

A narrow view of social contract theory frames the patentable subject matter question as a question of whether the economic benefits of innovation, prevention of free-riding, and incentives to invest outweigh the increased prices and exclusion of competition. The danger of assessing the cost on a purely economic basis in these circumstances is that the costs of patenting might be social, ethical, religious, environmental or scientific, and do not fit neatly into an economic

---

32 James Boyle, "Enclosing the Genome: What the Squabbles over Genetic Patents Could Teach Us" in F. Scott Kieff, *Perspectives on the Human Genome Project* (Academic Press, 2003) 97 at 109.

33 Advisory Council on Intellectual Property, *Patentable Subject Matter: Issues Paper* (2008) <http://www.acip.gov.au/library/Patentable%20Subject%20Matter%20Issues%20Paper.pdf> (accessed 13 November 2008) at 1.

34 This line of thinking emerged *Liardet v Johnson* (1778) 1 Carp Pat Cas 35 (NP), which characterised the grant of a patent, not as a privilege, but as a *quid pro quo*, wherein a grant of monopoly was given in exchange for the proper disclosure of the working of the invention. See also *Turner v Winter* (1787) 19 Eng Rep 1276 where the Court noted that "[t]he consideration, which the patentee gives for his monopoly, is the benefit which the public are to derive from his invention after his patent is expired." However, economic rationales for patent law began to dominate as the eighteenth century progressed. See Edward C. Walterscheid "The Early Evolution of the United States Patent Law: Antecedents (Part 4)" (1996) 78 Journal of the Patent and Trademark Office Society 77 at 104–106. Similarly see the second reading speech for the *Patents Act 1990* Australia: "[t]he essence of the patent system is to encourage entrepreneurs to develop and commercialise new technology.": Commonwealth of Australia, "Patents Bill 1990: Second Reading" Senate, 29 May, 1990 <http://parlinfoweb.aph.gov.au/PIWeb/view_document.aspx?id=562046&table=HANSARDS> (2 November 2004). A different theory underlies the US system, namely patents as a reward of inventors: Edward C. Walterscheid "Patents and Manufacturing in the Early Republic" (1998) 80 *Journal of the Patent and Trademark office Society* 855 at 856.

35 Adam Smith, *An Inquiry into the Nature and Causes of the Wealth of Nations* (5th ed, Methuen, 1904) at 159–160.

analysis. The dangers of overstating patent law in purely economic terms can be stated thus:

> Economics can be useful when it is viewed as a science which examines the decision making process, the study of optimization behaviours subject to constraints. Economics cannot be helpful if it is viewed as a precise tool that can mechanically and independently determine the outcomes of complex problems. ... [I]t can be used to assist in the framing of issues and in isolating the appropriate factors for judicial consideration. Economics is not helpful, however, in the inherently subjective process of weighing and quantifying competing concerns. It is wrong not to recognize this limitation, and it is dangerous to assume that difficult, value-laden decisions ... can be decided mechanically by appealing to an economic formula.[36]

Such a question properly recognises that at times questions of patentability cover social, ethical, religious, environmental, and scientific issues, in addition to economic and legal ones. The latter by themselves will sometimes form an insufficient basis on which to ground patent law. What is sought to be achieved in the course of this argument then, is not to displace economics as a tool by which to assess patent law. It is to develop an alternative analytical framework by which to assess patentable subject matter, and one which at the very least leaves the door open to these broader considerations. At the same time, it is recognised that any such analysis needs to be consistent with the principles developed over patent law's very long history.

The final aim of this book is to bring some clarity to the nature of software. The nature of software as both a description of a computable process, and the mechanism by which that process is actually performed, is apt to cause confusion. For example, it may be difficult for some to understand how something which can be bought, installed on a computer and then run, causing certain images to appear on a screen, might not in fact be a physical artefact at all. By applying the analytical framework which will be developed in this book to the special case of software, it is hoped that some light can be shed on the nature of both software development, and software itself, and avoid further mischaracterisation in the future.

2　Outline

Chapter 1 explores important introductory material, namely the nature of software, although many of the ideas explored are developed in more detail in

---

36　Peter J. Hammer, "Free Speech and the 'Acid Bath': An Evaluation and Critique of Judge Richard Posner's Economic Interpretation of the First Amendment" (1988) 87(2) *Michigan Law Review* 499 at 499.

Chapter 5. In particular, three aspects of software are identified as key to a proper understanding of it. First, successive generations of software have been freed from specific hardware limitations through a process of abstraction. Understanding the level of abstraction means understanding the connection between the software and the physical constraints of the computer. The level of abstractness varies depending on the task at hand, and therefore analysis can be context-specific. Second, software development depends on reuse, with software components being built using layers, libraries, frameworks, and design patterns. Reuse is a key characteristic of the software industry, and one without which it could not continue. Finally, software entities are highly complex, typically involving large numbers of unique components. The scale of complexity has been said to be similar in magnitude to that of an aircraft, but the unique nature of each component means there are no economies of scale. Chapter 1 also explores the nature of the relationship between software and mathematics, and explains how it is that software is said to be identical with, or at least isomorphic to, mathematics at both formal and structural levels.

Chapter 2 will explain in detail why the patenting of software is problematic from both practical and theoretical perspectives, tracing these difficulties to the three key features of software identified in Chapter 1. The chapter will also document the way in which the courts of the US, the EU, and Australia have attempted to deal with these problems, and describe why it is that the software patent issue should be dealt with as a patentable subject matter issue rather than as an issue of novelty or non-obviousness. It will be shown why the current approaches are insufficient, and to introduce the notion that a fresh approach to the software patent problem is needed.

In particular, it will be argued that the problem lies in a failure to address the limitations of a narrow economic (or utilitarian) approach to patent law which does not permit consideration of wider issues. Human rights jurisprudence will be introduced as a relevant consideration, setting up the next chapter which will demonstrate how the unpatentability of mathematics can be accounted for on the basis of free speech/thought.

The ultimate aim of the book is to shed fresh light on the scope of patentable subject matter, to clarify the nature of software, and by doing so to explain why software should not be patentable subject matter. Chapter 3 begins that journey by building on the isomorphism between software and mathematics identified in Chapter 1. The chapter first explores the history and philosophies of mathematics in an attempt to ascertain its nature. It will then be shown that legal accounts of the nature of mathematics are insufficient, as they cannot be reconciled with the contributions of the various philosophies of mathematics to a full understanding of what the nature of mathematics is.

Chapter 4 further explores the nature of mathematics, with a view to constructing a more suitable explanation of why mathematics is non-patentable. To do so, a different approach is taken. Whereas the historical, philosophical and legal accounts of the nature of mathematics have focused on what mathematics *is*, it will be shown that the better focus is on what mathematics *requires* for its further progress. By answering that question, it will be possible to avoid the impossibility of providing a definitive unifying account of what mathematics is, and to answer the question of whether mathematics is a proper subject for patent protection. It will be shown that the underlying reason for excepting mathematics from patentability is that mathematical innovation depends on freedom of thought and expression.

Rather than seeking to recharacterise patent law as the competition between inconsistent human rights, it will, however, be shown how such freedoms can be reconciled with a traditional understanding of the scope of patent law. The mechanism by which that reconciliation is to be achieved is the distinction between the non-patentable fine arts and the patentable useful arts. It will be argued that the concept of the useful arts, or technology, as it might be known in contemporary language, provides a suitable way of distinguishing between invention and other intellectual pursuits. The analysis of the way freedom operates with mathematics, combined with a foray into the philosophy of technology, leads to the derivation of a three-dimensional analytical framework by which it is possible to distinguish between the fine arts and the useful arts in a structured fashion. The analysis is then applied to mathematics to show why mathematics is a fine art and not a useful art, notwithstanding its role in many technological pursuits, because it is intangible rather than physical, expressive rather than functional, and aesthetic rather than rational.

Chapter 5 applies the anlaytical framework developed in Chapter 4 to software, exploring the expressive, aesthetic, and intangible nature of software development to support a contention that software is also a fine art. In particular, it will be suggested that the interrelationship of software and hardware, referred to as abstractness in Chapter 1, acts as an important determinant. Where software and hardware are closely interrelated, the physical limitations may constrain the software such that expressive and aesthetic considerations are similarly limited. Where software is merely ancillary to a physical device, this does not change the nature of programming, but it may change what is sought to be patented, namely a physical device in which software is merely a subsidiary component. Determining whether it is software, or a physical device which is actually being claimed requires difficult determinations of fact in any particular case, but the difficulty of the choice does not obviate the need to draw such distinctions.

Having demonstrated why programming, like mathematics, is not a useful art, Chapter 6 explores the implications of that proposition. The value of the analytical framework developed, and the current state of patent law in Australia, the UK, and the US are assessed. The importance of characterisation is also acknowledged. Next, the role of subject matter is considered. The need to exclude software from the patent paradigm suggests that subject matter is not a "failed gatekeeper", but an important mechanism by which "soft" issues such as ethics and social considerations, can be reconciled within patent law. Finally, some remarks are addressed to a broader issue. Having suggested that software ought not to be protected by patent law, it will be also be submitted that there is no need to warp patent law to protect the software industry. Alternative protection mechanisms are explored and found to be better suited to crafting an Information Age response to an Information Age problem.

3 Other matters

The law stated herein is correct up to 26 June 2018.

1 The nature of software

1 Introduction

To understand why software is an issue for patent law, it is first necessary to understand a number of things about how modern software is organised and how any particular software component is likely to be built. The nature of software will be returned to in Chapter 5 in much greater detail. However, an understanding of the software development process is necessary in order to properly understand the problems which attend on the software patent cases discussed in Chapter 2.

Through an exploration of the nature of software, four things will become clear. First, the process of abstraction away from specific hardware limitations has been the key to the success of software. Second, the way in which software systems are constructed using layers, libraries, frameworks, and design patterns makes it clear that one important goal of software development is to encourage reuse. Third, the development of a particular piece of code tends to be very context-specific – not all software development activities are alike. Finally, software entities tend to be complex, typically involving large numbers of unique components. Abstraction, re-use, context, and complexity all raise important considerations for the patentable subject matter inquiry which will be developed in later chapters.

This chapter will also introduce one of the major themes of the book – that software is isomorphic to mathematics. This isomorphism can be demonstrated both at the formal level and at the structural level. Whilst there are undoubtedly some differences between the two, the *prima facie* identity established provides a proper basis for exploring software's patentability by reference to the patentability of mathematics.

2 What is software?

Early in the history of computing, collections of physical components which make up a computer began to be called "hardware". Software was then coined to refer to the non-physical aspects of the computer, especially the programs, or sets of instructions "which cause a computer to perform a desired operation or series of operations".[1] Such a definition is useful to a certain extent, but overlooks the complex relationship of hardware and software, particularly when an externally stored program[2] is loaded into the memory of a computer. It also encourages a focus on the end product of the software development process – executable code – at the expense of other artefacts which exist at intermediary stages; and which in some sense determines the final product.

So to truly understand the nature of software, it is necessary to understand how the development of software has evolved since the early days, as much of this evolution has determined the way in which software developers currently operate. It is then possible to see how software developers take an abstract idea and develop it into working software through the paradigm known as top-down programming. It will also be possible to make some general observations about what makes software unique.

2.1 The evolution of modern software

A Machine code

Machine code, or native code, is the medium in which all software was originally written, and the form in which software ultimately executes on a computer. Machine code is a string of 1s and 0s (called bits) of a set length which tell the computer's central processing unit (CPU) which instruction to perform, where to get the data and where to put the result. An example of a 32-bit instruction is given below:[3]

Table 1.1 Example machine code instruction

1	0	0	0	0	0	0	0	0	1	0	0	0	0	0	1	0	0	0	0	1	1	0	0	0	0	0	0	0	0	0	0
Instruction (ADD)				Output (L4)				Input 1 (L2)				Input 2 (L3)				(unused)															

This instruction tells the CPU to **add** the contents of memory location 2 to memory location 3 and put the result in memory location 4. The available instructions set varies betweeen processors, but typically includes such operations as:

- arithmetic operations, such as **add** and **subtract**;
- logic instructions, such as **and, or** and **not**;[4]

---

1 "Software" in *The Macquarie Dictionary,* (4th ed, The Macquarie Library Pty Ltd, 2003).
2 Such as on punched cards, a floppy disk, or in more recent times, a CD or DVD.
3 Adapted from Arvind Asanovic, "Early Developments: From the Difference Engine to IBM 701" (Lecture Slides, MIT, 2005) <http://ocw.mit.edu/courses/electrical-engineering-and-computer-science/6-823-computer-system-architecture-fall-2005/lecture-notes/l01_earlydev.pdf> (11 September 2011).
4 For example, the logical operator "and", is written "∧". a ∧ b evaluates to *true* where both *a* and b are both *true*, but otherwise evaluates to *false*. In binary, true is represented by the number 1 and false by the number 0. So, 1 ∧ 1 = 1, but 1 ∧ 0 = 0. This might be done

- data instructions, such as **move**, **input**, **output**, **load**, **store**;[5] and
- control flow instructions, such as **goto, if ... goto, call**, and **return**.

Each instruction has its own corresponding binary number. This can make writing programs in machine code tedious and time-consuming. Errors are easy to make and hard to find. Further, the mappings of instruction to number can vary from processor to processor. This means there is little portability between machines.[6]

So early on, programmers began to look for ways of moving towards more human-readable representations of computer instructions.

B   Assembly language

The next development was to write a program which could take care of converting human-readable instructions to number strings, called an assembler/disassembler. The assembly language version of the machine code instruction above looks like this:

ADD (L4, L2, L3)

"ADD" is obviously more memorable than 100000, but there is still a one-to-one correspondence between assembler instructions and machine code instructions. Seemingly simple operations require a number of machine-level instructions,[7] and assembly programmers still have to work within the constraints of hardware-specific instruction sets.[8] So they set to work on hiding the low-level details of the computer in another way.

C   High-level languages

Most programming these days is done using high-level languages, which are another step closer to natural language. High-level languages also hide the

---

for each bit (column) of the data in L2 and L3 in the example above. If so, the value to store in L4 = 00010 ∧ 00011 = 00010.

5 These commands are used for interacting with data available in various parts of the computer, including the CPU, hard disk, RAM, displays, input devices.

6 This was a particularly big issue early in the history of computing, where there was no standardisation of architecture. Although it may appear to be less of an issue now, where most computers are based on the Intel Pentium architecture, there are still many alternative architectures in use, especially for smaller electronic devices like mobile phones.

7 Assembler code for multiplying two numbers on an AVR chipset requires no less than 30 separate instructions. See Gerhard Schmidt, "Mult8.asm" on *Tutorial for learning assembly language for the AVR-Single-Chip-Processors* <http://www.avr-asm-tutorial.net/avr_en/calc/MULT8E.html> (24 July 2008).

8 Despite the difficulties, assembly language is still used today in applications where efficient performance in the software is valued over the manual labour involved in creating it, for example in device drivers, embedded systems, and computer game routines. See Wikipedia, "Assembly language" <http://en.wikipedia.org/wiki/Assembly_language> (18 August 2008).

machine architecture from the programmer. A high-level representation of the assembly code above would be:

```
a = b + c
```

When using high-level languages, programmers no longer have to worry about memory locations, architecture-specific instructions, or other low-level hardware details. In order to be executed by the computer, however, high-level language commands need to be translated into machine code. As such, high-level languages are often classified according to how this translation is achieved.

D Compilers

Compiled languages cannot be run directly on a computer in their original (source code) form. Before they can be executed, a translation program (compiler) translates the high-level language instructions to machine-specific instructions (the result of compilation being known as object code). Common compiled languages include C and Java.

E Interpreters

Interpreted languages do not have this preparatory step. A program called an interpreter translates high-level code to machine code as the code is run. Modern examples of interpreted languages include Python, Perl, PHP and Ruby. Since no compilation is involved, the development and testing of software written in these programs is easier. The downside of interpreted languages is that they generally run slower and require more memory.

F Software stacks

But the abstraction of the software development process does not end with natural languages. A different type of abstraction is achieved by breaking software down into a series of layers, called a stack, where each layer offers the layer above access to its services, but only via a set of higher-level functions. The typical layers involved in the operation of a modern computer are discussed below.

Firmware, kernels, and operating systems: Firmware exists right at the borderline between hardware and software, and can be thought of as a computer program embedded in a hardware device. The most familiar role for firmware in modern computers is as the Basic Input Output System (or BIOS) which launches a computer's startup process by detecting the installed hardware elements, then loading the operating system.

Operating just above the level of the firmware is an operating system's kernel. The kernel handles the lowest level of interactions with the hardware of a computer. The most commonly known kernel is Linux, the kernel of a GNU/Linux operating system. The job of the kernel is to "manage the computer's resources and allow other programs to run and use these resources".[9] These resources include the CPU, memory (RAM) and various input/output devices such as displays, disk drives, mice, and keyboards. Since there is usually more than one program running at any time on a computer, the kernel can be thought of as deciding which program gets access to what, when, and for how long. The kernel also takes care of communication between programs.

Other parts of the operating system perform a wider range of higher-level services, including:[10]

- program execution (starting and stopping applications);
- responding to events (e.g. "somebody hit the "A" key – send an A to the current program!");
- memory management (e.g. "Firefox opened a new window – give it more memory!");
- multitasking (e.g. that which allows you to surf the web, run a mail merge, and print at the same time);
- disk access and file systems;
- device drivers (e.g. making sure Comic Sans MS looks the just as ugly on the printer at work as it does on the printer at home);
- networking (e.g. sending data from your computer to the email server in bite-size chunks, and making sure it all got there);
- security (e.g. who can log in? who can see this file?); and
- graphical user interfaces (windows, buttons, menus, and so on).

Libraries, frameworks, and patterns: Software developers usually look to break software down into sets of components which can be re-used beyond the problem at hand. As such, most, if not all, applications are built using a series of independent components, which are often written by third-party developers, or as a collaborative effort between various parties.

A library (or module) is a "a substantial number of computer program modules designed to solve a wide range of problems in [a] given area".[11] For example, a library might contain a function for calculating mathematical or scientific formulae,[12] or functions for managing files.[13] Libraries allow

9 Wikipedia, "Kernel (computer science)" <http://en.wikipedia.org/wiki/Kernel_(computer_science)> (23 July 2008).

10 This list was adapted from Wikipedia, "Operating system" <http://en.wikipedia.org/wiki/Operating_system> (23 July 2008).

11 "Software Libraries, Numerical and Statistical" in Anthony Ralston, Edwin D. Reilly and David Hemmendinger (eds), *Encyclopedia of Computer Science* (4th ed, Wiley, 2003).

12 For the C language, common mathematical functions are available in the "math" module, distributed with the compiler. A list of numerical and scientific libraries (or modules as they are called in Python) are available at <http://wiki.python.org/moin/NumericAndScientific>> (29 July 2008). For the Perl language, see <http://cpan.org>.

13 For example, "io.h" in C, "sys" in Python, or "csv" in Python for interpreting comma separated value files.

programming effort to be concentrated on efforts that can be re-used. Libraries can be thought of as extending a language.

A framework is "the skeleton of an application that can be customized by an application developer"[14] and provides a set of reusable components such as libraries, as well as a design or design or pattern by which to combine them to form a whole application. Frameworks consist of both code, in the form of components, as well as a method of organising those components into a larger whole. As an organising concept, they begin to stray beyond the realm of software as "code" into the realm of the abstract.

Patterns, or design patterns, are a "general reusable solution to a commonly occurring problem in software design".[15] The original concept of patterns was imported to the software field from architecture,[16] and allows a design to be built from a catalogue of generalised solutions to recurring problems. A design pattern "describes the problem, the solution, when to apply the solution, and its consequences".[17] Since the mid-1990s,[18] design patterns have had an increasingly influential role in the development of software.

Patterns arguably exist at such a high level of abstraction that one might wonder whether they are software at all. They are certainly a useful design tool and also provide a vocabulary for describing the way software works. Intuitively speaking, though, unless they are realised in some executable form,[19] they are not software.

Scripting languages and domain-specific languages: Applications are sometimes developed in such a way that they can be extended by plug-in modules. Scripting languages are mini-languages designed for extending such applications. They allow users to write "programs within a program", to automate repetitive tasks within the program. A common example would be Microsoft Word macros, written in Visual Basic.[20]

Similarly, domain-specific languages solve needs specific to a particular problem domain. For example, Structured Query Language, or SQL, is the language used to select, filter, order, and run calculations on data stored within

---

14 Ralph E. Johnson, "Frameworks = (components + patterns)" (1997) 40(10) *Communications of the ACM* 39 at 39.
15 Wikipedia, "Design pattern (computer science)" <http://en.wikipedia.org/wiki/Design_pattern_(computer_science)> (29 July 2008).
16 See Christopher Alexander et al., *A Pattern Language: Towns, Buildings, Construction* (Oxford University Press, 1977).
17 Erich Gamma et al., *Design Patterns: Elements of Reusable Object-Oriented Software* (Addison-Wesley Professional, 1994).
18 The take-off in design patterns is generally attributed to the publication of the so-called "Gang of Four" book, Gamma et al., above n 17.
19 By this it is meant that they should be manifested in either machine code, assembly language or a compilable or interpreted high-level language.
20 For a list of examples, see Matt Reich "10 Awesome Uses for Automator Explained", *EnvatoTuts+*, 5 January 2011 <https://computers.tutsplus.com/tutorials/10-awesome-uses-for-automator-explained–mac-15845> (27 February 2017).

databases.[21] Cascading Style Sheets are another example, and are widely used to manage the formatting of web sites (font size, background colours, and so on).[22]

2.2   The modern software development process

It is also important to understand the process of software development, being "one of gradual refinement and elaboration"[23] from idea to source code. Ogilvie identifies six individual levels of abstraction within the software development cycle,[24] and whilst many of them may take place in parallel, they are helpful in explaining how software developers 'bridge the gap' from concept to code.[25]

A   *Main purpose*

This is a high-level description of the problem which the software is intended to solve.

B   *System architecture*

As the programmer learns more about the problem space, they will begin to break it into a series of smaller problems, or modules. "Each module performs a significant portion of the program's main purpose and is eventually implemented as a distinct section of the source code".[26] The system architecture describes how the program is dissected into these modules, and the interactions and/or relationships between them. This can be described by the way they 'call' each other (control flow); the way they push data to each other (data flow); or by the way they are 'nested' inside each other.[27]

C   *Abstract data types (ADTs)*

Abstract data types, also known as objects, define the types of data stored and the operations which can be performed on each data type.[28]

---

21 For an introduction see John Worsley and Joshua Drake "Chapter 3: Understanding SQL" in *Practical PostgreSQL* (O'Reilly, 2001) <http://www.faqs.org/docs/ppbook/c1164. htm> (3 November 2008).

22 For an introduction see Dave Raggett "Introduction to CSS" <http://www.w3.org/ MarkUp/Guide/Style> (4 November 2008).

23 Peter G. Spivack "Comment: Does Form Follow Function? The Idea/Expression Dichotomy in Copyright Protection of Computer Software" [1988] 35 *UCLA Law Review* 723 at 729.

24 John W.L. Ogilvie, "Defining Computer Program Parts Under Learned Hand's Abstractions Test in Software Copyright Infringement Cases" (1992) 91 *Michigan Law Review* 526.

25 It should be noted that the process involved describes the modern paradigm of object-oriented programming.

26 Ogilvie, above n 24 at 534.

27 To make an analogy, a 'family' module might have two 'parent' modules and two 'children' modules.

28 Ogilvie, above n 24 at 536.

D *Algorithms and data structures*

Algorithms and data structures are more computer-specific representations of operations and data types than those above. What distinguishes algorithms and data structures from ADTs is that there may be a number of specific algorithm and data structure combinations that correctly implement the requirements of the ADT. "Although data structures and algorithms are more specific than ADTs and may even depend on certain programming language features for use, they are independent of any specific programming language or piece of literal source code."[29]

Algorithms, also known as procedures, are a 'unit' of computation which perform a specific task. Whilst an "operation merely identifies a desired result, an algorithm specifies every step necessary to accomplish that result".[30] Typically, procedures take data as an input, and act on it to create a new set of data as an output. For example, the following represents a procedure for sorting a set of numbers from lowest to highest:

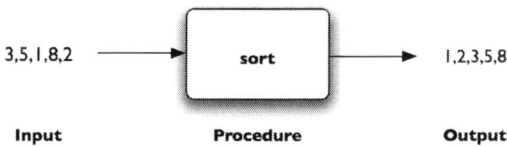

| Input | Procedure | Output |

Figure 1.1 Sort procedure

Procedures are usually only considered at this somewhat abstract level, so as not to be constrained by the specific limitations of either computer hardware and a particular programming language, although their operation may be described in more detail than here, either in natural language, a flowchart, or in some sort of pseudo-code.

Data structures: describe how a computer is to represent the data in a data type. A data structure "consists of one or more variables of the basic data types, which are organized in some specified combination of arrays, records, and pointers".[31] Data structures are defined at an abstract level which does not require any consideration of the way they will be represented in the memory of a computer. Considered together, procedures and data structures are a preliminary sketch of a program and are usually written in natural language, or formal notation language.

29 Ogilvie, above n 24 at 540.
30 Ogilvie, above n 24 at 536.
31 Ogilvie, above n 24 at 539.

E   Source code

With a preliminary design in place, the final step is for the procedures and data structures to be implemented in the programmer's programming language of choice. The programmer is limited in the expression of the code at this stage by the syntax and grammar of the language. Depending on context and choice of language, the hardware on which the program is to run may also influence the way the source code is written. Because it is in many ways the final stage of the development process, the software reflects all the previously discussed levels of abstraction. As Ogilvie notes:

> A program's entire range of abstraction is embedded in its code in roughly the same way a novel's characters and plot are embedded in its text. Therefore, care must be taken to avoid confusing code as the embodiment of parts at every level of abstraction with code as a level of abstraction in its own right.[32]

F   Object code

This is the final, executable form of software. The translation from source code to object code is usually automated by a compiler or interpreter, although it can involve some degree of manual configuration by the programmer. As noted above in relation to interpreted languages, the distinction between source code and object code can be very subtle indeed.

G   Putting it all together

What the discussion thus far should make clear is that the development of software products is dependent on a series of factors. These will now be discussed in turn.

Abstractness: The history of computer science shows consistent movement away from hardware towards the abstract realm of ideas. Assembly language is an abstraction of machine code, and high-level languages are a further abstraction of assembly language. Each of the layers in the development stack above can also be seen as an abstraction away from the specifics of the layer below.

Modern software is largely "pure thought-stuff, infinitely malleable".[33] Not only can it be changed, it is also "embedded in a cultural matrix of applications, users, laws, and machine vehicles [which] all change continually, and their changes inexorably force change upon the software product".[34]

Further, different levels of abstraction may sit on different conceptual layers. On one conceptual level, the software can be considered to be a description of the functionality that the program will have when it is executed. At the same

---

32 Ogilvie, above n 24 at 541.
33 Frederick P. Brooks Jr, "No Silver Bullet: Essence and Accidents of Software Engineering" (1987) 20(4) *Computer* 10 at 12.
34 Brooks, above n 33 at 12.

time, the software *causes* the executing computer to exhibit that behaviour. These conceptual layers are reflected to some extent in compiled languages, where the source code is descriptive and the object code is functional. But such neat divisions can be misleading. Even machine code is descriptive to a programmer who understands it. With interpreted languages, the source code *is* the object code, and is thus both descriptive and functional at the same time.

Software is in essence invisible, or intangible. Unlike other abstractions, such as a floor plan of a building, software abstractions are "not inherently embedded in space. Hence, [they have] no ready geometric representation in the way that land has maps, silicon chips have diagrams, and computers have connectivity schematics."[35] Various diagrams are used to attempt to represent software,[36] however these fail to capture the whole of what software is.

Software is not subject to myriad physical uncertainties[37] which otherwise have to be overcome as a part of the experimentation process, since they are "largely determined, in advance, by the specification, the flow chart, the rules of the programming language, the programming conventions, and the dictates of logic and mathematics".[38]

As a result, context is critical when considering software. The context in which a software component is located is key to determining how much it has to do with the hardware on which it runs and the dependence which that component has on underlying limitations of the stack upon which it is built. For example, writing firmware components in assembler is a task which is largely constrained by the hardware configuration that it is to be written for, and will probably rely less on reusable components than will writing a macro for a word processor.

Re-use: The development of software from machine code up to high-level languages demonstrates how progress in software depends on re-use of the work of the past – assemblers were written in machine code, compilers were written in assembly language, and so on. It is through the process of abstraction from the specific to the general that allows for re-use.

Re-use also influences the design of software by encouraging conformity. As a corollary, the expectation of conformity from users also encourages re-use. This expectation is often formalised into the notion of a standard.[39] Standards exist

---

35 Brooks, above n 33 at 12.
36 For example, flow of control, the flow of data, patterns of dependency, time sequence, and name-space relationships. See Brooks, above n 33 at 12.
37 See Jay Dratler Jr, "Does Lord Darcy Yet Live? The Case Against Software and Business Method Patents" (2003) 43 *Santa Clara Law Review* 823 at 854, and also Chapter 6 at 157–158.
38 Dratler, above n 37 at 855.
39 Standards can be developed privately or unilaterally, and may become a standard through custom or convention (a de facto standard, such as the Microsoft Word document format) by mandate (such as when a government body demands compliance with a certain standard) or formal consensus (such as through the formal vote of a standards organisations like the International Standards Organisation). They often take the form of "a formal document that establishes uniform engineering or technical criteria, methods, processes and practices": Wikipedia, "Standard" *Wikipedia.org* <http://en.wikipedia.org/wiki/Standard> (3 October 2008).

for such things as file formats,[40] user interfaces,[41] and network communication protocols.[42]

Complexity: Another factor making software unique is its complexity. As Brooks notes, "[s]oftware entities are more complex for their size than perhaps any other human construct because no two parts are alike. If they are, we make the two similar parts into a subroutine."[43] The number of lines of code in a program can provide a rough estimate of the magnitude of complexity of software products. For example, version 2.6.24 of the Linux kernel, which alone comprises only a small part of a GNU/Linux operating system, contained over 8.5 million lines of code.[44] Red Hat 7.1, a GNU/Linux distribution released in 2001, contained around 30 million lines of code.[45] Windows XP, released in October 2001, contained around 40 million lines of code.[46] A million lines of code is about 18,000 pages of printed text.[47]

On the other hand, there is a simplicity about software which is further unlike physical objects.[48] Software components are stable and predictable and lacking in uncertainty. Their behaviour can be predicted beforehand with complete accuracy, unlike for example, the momentum and location of an electron in orbit around an atom.[49]

---

40 Aside from the de facto standard of the Word document noted above, the alternative Open-Document Format was passed by the International Standards Organisation in 2006. See "Open Document Format for Office Applications (OpenDocument) v1.0" ISO/IEC 26300:2006 <http://www.iso.org/iso/iso_catalogue/catalogue_tc/catalogue_detail. htm?csnumber=43485> (3 October 2008).

41 See for example Apple Inc, "Introduction to Apple Human Interface Guidelines" [2008] *Apple Developer Connection* <http://developer.apple.com/documentation/UserExperi ence/Conceptual/AppleHIGuidelines/XHIGIntro/chapter_1_section_1.html> (2 October 2008).

42 The protocols used for the Internet, commonly referred to as TCP/IP were made the standard for all military computer networking by the US Department of Defense in March 1982. See Wikipedia, "TCP/IP" <http://en.wikipedia.org/wiki/TCP/IP> (8 October 2008). The official specification is documented in the following document: Internet Engineering Task Force, "Request for Comments 1122: Requirements for Internet Hosts" <http:// tools.ietf.org/html/rfc1122> (8 October 2008).

43 Brooks, above n 33 at 12.

44 Greg Kroah-Hartman, Jonathan Corbet and Amanda McPherson "Linux Kernel Development (April 2008)" [2008] *The Linux Foundation* <http://www.linuxfoundation.org/ publications/linuxkerneldevelopment.php> (28 August 2008).

45 David Wheeler, "Counting Source Lines of Code (SLOC)" <http://www.dwheeler.com/ sloc/> (28 August 2008).

46 Larry O'Brien, "How Many Lines of Code in Windows?" on *Knowing.NET* <http://www. knowing.net/PermaLink,guid,c4bdc793-bbcf-4fff-8167-3eb1f4f4ef99.aspx> (28 August 2008).

47 Codebases: Million lines of code, *Information is Beautiful*, <http://www.informationis-beautiful.net/visualizations/million-lines-of-code/> (14 October 2017).

48 This point is aptly illustrated by Dratler, above n 37 at 864–9.

49 So says Heisenberg's uncertainty principle, which lies at the heart of quantum physics. See Jan Hilgevoord and Jos Uffink "The Uncertainty Principle" *Stanford Encyclopedia of Philosophy* <http://plato.stanford.edu/entries/qt-uncertainty/> (21 October 2008).

These are the dimensions of software seen from the perspective of computer science and software development. But this is not the whole story. Hiding within the story of the history of software is an important relationship. The lowest level of computing, machine code, is pure number crunching. Computer instructions, and the data on which they act are represented as a series of numbers – 1s and 0s. Each processor instructions is a method of transforming these numbers. Numbers and their transformations are clearly the domain of mathematics, which suggests at least a similarity, and at most an identity. So are mathematics and software wholly identical? If not, what are the dimensions of their relationship?

3 How is software related to mathematics?

Most attention given by patent law scholars to understanding the relationship between mathematics and software has focused on the patentability of algorithms. Whilst an algorithm can be defined broadly to include any conceivable process, courts have called on the notion of a *mathematical* algorithm, which has been defined as "a procedure for solving a given type of mathematical problem".[50] Since algorithms are only really one component of software, however, such an understanding is an insufficient basis on which to build an understanding of the relationship.

There are two reasons why the algorithm has taken centre stage: one historical, one practical. Historically, software was written as one long sequence of code, or one algorithm. This remained the dominant approach until the late 1980s, when the top-down methodology took hold. All of the US Supreme Court cases on the patentability of software up to *Diehr* took place before this point. But this justification only goes so far, as there have been a number of cases since, none of which has taken this transformation into account. Further, the same courts when considering copyright have demonstrated a clear understanding of that change.[51]

Practically then, algorithms are important because they contain the function, as opposed to the structure, of software. Since functionality is what the patent regime tries to protect, it is to be expected that courts have kept algorithms at the centre of their focus.[52] The organisation of code, although important, is akin to a "mere scheme or plan",[53] and would be excluded by patent law.[54]

---

50 *Gottschalk v Benson* 409 US 63 (US Supreme Court, 1972); *IBM v Commissioner of Patents* (1991) 33 FCR 218 (Federal Court of Australia) at 220.

51 See for example the progression from the approach of the court in *Whelan Associates v Jaslow Dental Lab Inc* 797 F.2d 1222 (1986) to that in *Computer Associates International, Inc. v Altai, Inc.* 982 F.2d 693 (1992).

52 It should be noted however that data structures have sought to be protected as well. See for example *In re Alappat* 33 F.3d 1526 (1994); *Welcome Real-Time v Catuity* (2001) 113 FCR 110.

53 *Cooper's Application* (1902) 19 RPC 53.

54 Alternatively, it may be considered part of the non-patentable idea rather than a substantive part of the invention. Finally, the organisation of the code may be too far removed from the "useful product" required by the court in *Grant v Commissioner of Patents* [2006] FCAFC 120.

Thus it is both sufficient and useful to consider the relationship of software and mathematics through the lens of the algorithm.

Algorithms, as a formalised description of a process, give a clue as to that relationship of software and mathematics. In the early part of the 20th century, the mathematician David Hilbert sought to prove that the truth of mathematical principles could be formally demonstrated.[55] A formal proof is in effect a formalised version of the mathematical algorithm as described above. A formalised notion of mathematics seeks to prove the truth of a mathematical theorem as follows:

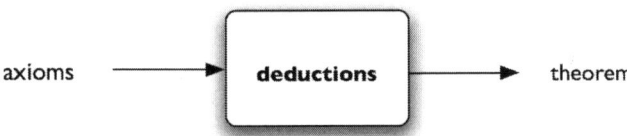

Figure 1.2 Deducing theorems from axioms

The formalist deductive process takes as its starting point a set of foundational truths, or axioms.[56] Each step of the deductive process involves the application of any one of a set of rules, whose truth can also be considered axiomatic. Although the formalist programme was ultimately unsuccessful, the notion of the reduction of reasoning to a series of mechanical steps led to a number of formal models of computation, which, despite their diversity, have been shown to be equivalent "in the sense that each analysis offered has been proved to pick out the same class of functions".[57] The most famous of these is the Turing machine. The formal equivalence of mathematics, the Turing machine and modern computers are considered below.

A   Formal equivalence: the Turing machine

The Turing machine was conceived by Alan Turing in 1936 as a formal statement of computability, or in other words, a definition of what could be computed. At the heart of this definition was a device called the Turing machine. A Turing machine consists of the following components:

---

55 Hilbert's formalist programme is discussed in greater detail in Chapter 4.
56 Exactly what is to be considered axiomatic can itself be a deep philosophical issue. This will be explored in more detail in Chapter 4. Russell and Whitehead used formalism in an attempt to deduce the whole of mathematics from the rules and axioms of logic in order to prove that logic was the proper foundation of mathematics.
57 Jack Copeland, "The Church-Turing Thesis" on *AlanTuring.net* (2000) <http://www.alanturing.net/turing_archive/pages/Reference%20Articles/The%20Turing-Church%20Thesis.html> (22 May 2007). See also Todd Rowland, "Church-Turing Thesis" *Wolfram MathWorld* (2002) <http://mathworld.wolfram.com/Church-TuringThesis.html> (accessed 23 April 2007) at 3. For a comprehensive survey, see Stephen C. Kleene *Introduction to Metamathematics* (North-Holland, 1980) at 12–13.

1   a control unit, which can assume any one of a finite number of possible states;
2   [an infinitely long] tape, marked off into discrete squares, each of which can store a single symbol taken from a finite set of possible symbols; and
3   a read-write head, which moves along the tape and transmits information to and from the control unit.[58]

On each iteration of a cycle, the Turing machine's next action will be determined by a table of "transition rules".[59] For example, in the image in Figure 1.3,[60] the symbol 'd' and the state 'q3' determine the next step taken by the machine. These actions might include:

•   writing or erasing the current symbol in the current position of the tape;
•   moving the tape left or right, or leaving the tape where it is; or
•   changing the internal control state.[61]

When the machine reaches a state for which there is no transition rule (or more than one), it is taken to have completed its computations, and the machine halts. An example of a transition table is given in Table 1.2:[62]

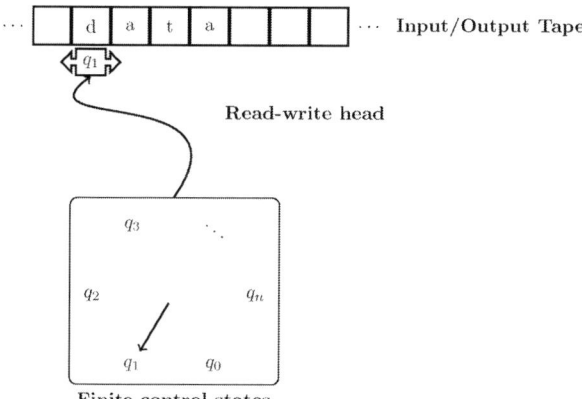

Figure 1.3 The Turing machine

58 "Turing Machine" in Anthony Ralston, Edwin D. Reilly and David Hemmendinger (eds), *Encyclopedia of Computer Science* (4th ed, Wiley, 2003).
59 David Barker-Plummer "Turing Machines" *Stanford Encyclopedia of Philosophy* <http://plato.stanford.edu/entries/turing-machine/> (19 July 2008).
60 Adapted from <http://www.texample.net/tikz/examples/turing-machine-2/> (16 October 2017).
61 Alan M. Turing, "On Computable Numbers, with an Application to the Entscheidungsproblem" (1937) 47 *Proceedings of the London Mathematical Society* 230.
62 Adapted from Turing, above n 61 at 233.

This example works as follows. This program endlessly loops through each of the control states b, c, e, and z in order, and can be represented diagrammatically as follows:

Table 1.2 A Turing machine transition table

Configuration		Behaviour	
Control state	Tape symbol	Tape operations	Final control state
b	blank	Print "0", Move right	c
c	blank	Move right	e
e	blank	Print "1", Move right	z
z	blank	Move right	b

The effect on the tape is to print alternating ones and zeros with a blank space in between, as shown in Table 1.3:

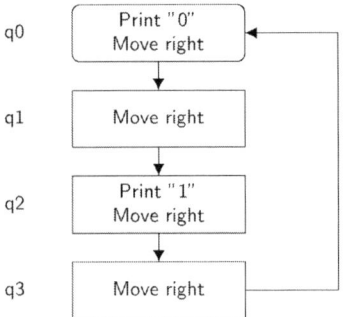

Figure 1.4 Turing machine control state flow diagram

Table 1.3 Turing machine output tape

0		1		0		1		0		1	. . .

Recalling the make-up of a typical machine-language instruction in Table 1.1, this gives a sense of how the Turing machine relates to modern computers. Despite the apparent simplicity of this machine, it is capable of much more complex calculations than the above example suggests. In fact, Turing proposed that "[Turing machines] can do anything that could be described as 'rule of thumb' or 'purely mechanical'."[63] In other words, "[w]henever there is an effective method for obtaining the values of a mathematical function, the function can be computed by a Turing machine."[64] This statement cannot be proved in a precise fashion because what is computable and what is not is something

---

63 Alan M. Turing, "Intelligent Machinery" in Jack Copeland (ed), *The essential Turing: seminal writings in computing, logic, philosophy* (Oxford University Press, 2004) at 414.
64 B. Jack Copeland, "The Church-Turing Thesis" *Stanford Encyclopedia of Philosophy* <http://plato.stanford.edu/entries/church-turing/> (20 July 2008).

which is largely intuitive. As such it is more of a "working hypothesis",[65] but nonetheless one which continues to be applied. For this reason, the group of functions which are computable are also known as functions that are "intuitively calculable" (or computable).

More importantly, all known calculable functions share a common feature. Alonzo Church was able to show that they have the quality of being general recursive.[66] Church was also able to demonstrate that any such algorithms could be written in a mathematical notation called the lambda calculus (λ-calculus). The λ-calculus and Turing machine are demonstrably "equivalent, in the sense that each picks out the same set of mathematical functions".[67] As such, the notion that all known effectively calculable mathematical algorithms can be calculated by a Turing machine is known as the Church-Turing thesis.

Although the focus of the Church-Turing thesis is on determining the boundaries of computability, the formal equivalence of mathematics and software follows from the equivalence of λ-calculus and Turing machine programs, in that all computable functions can be solved either by mathematical methods or by a Turing machine program.

But before a true equivalence can be reliably claimed, a further link must be made between real-world computers and the abstract definition of a Turing machine. If real computers and Turing machines are equivalent, then it follows that software and mathematics are identical, because anything that can be written in software can be described in the λ-calculus.

B   Are computers and the Turing machine equivalent?

The Strong Church-Turing thesis holds that a Turing machine can do whatever a computer can do. Support for this proposition is to be found in similarities between the λ-calculus and programming languages. For example, the λ-calculus has been referred to as the smallest universal programming language.[68] The λ-calculus has also infliuenced the design of a number of functional programming languages such as Lisp, ISWIM, and ML.[69]

65 Emil L. Post, "Finite Combinatory Processes – Formulation 1" (1936) 1 *Journal of Symbolic Logic* 103, cited in Copeland, above n 64.
66 A simple example of a partially recursive function is the Fibonacci number series (1, 2, 3, 5, 8, …), where the next number in the series is the sum of the previous two numbers in the series. Although the series is defined by reference to itself (this is what makes it recursive) we know that working out. say, the 15th element in the series can be broken down into a fixed sequence of steps. A general recursive function is also defined by reference to itself, but in such a way that it creates an infinite loop. The classic example of this is Epimenides' liar paradox, which can be put in the form "This phrase is false".
67 Copeland, above n 64.
68 For a thorough account of the use of the λ-calculus as a programming language itself, see John R. Harrison "Chapter 3 – Lambda calculus as a programming language" *Introduction to Functional Programming* (Lecture Notes, Cambridge University, 1997) <http://www.cl.cam.ac.uk/teaching/Lectures/funprog-jrh-1996/> (5 November 2008).
69 See Harrison, above n 68 at 25–26.

One difficulty in accepting the equivalence of Turing machines and computers is the Turing machine's unlimited storage. Unlimited storage is an important feature of the Turing machine because it "ensures that no computable function will fail to be Turing-computable solely because there is insufficient time or memory to complete the computation".[70] As a practical matter, the disparity falls away if it is assumed that the storage space of a real computer can be infinitely extended if required.[71] Alternatively, one might prove that anything calculable by a real-world computer was at the very least a subset of the functions effectively calculable by a Turing machine. If this was so, it would mean that the overlap between mathematics and real-world computers was a complete one.

Another major difference between the Turing machine and modern computers is the way in which data is accessed. The Turing machine is a sequential-access machine, since it processes the memory (tape) one square at a time. Modern computer memory is not sequential. Any position in memory can be accessed at any time. A computer based on such a design is called a random access machine (RAM). Despite these design differences, however, it has been shown that both designs are of equivalent power.

Since a Turing machine can simulate a RAM, and a RAM can simulate a Turing machine,[72] it follows that anything which can be designed for one machine can be run on the other (via the simulator, if not directly). Since these machine-level simulations of a modern computer are possible on a modern computer, our discussion of the development of assembly and higher-level languages above makes it clear that software written in modern computer languages can be made to run on a Turing machine and vice versa. Put simply, this means that all modern software is equivalent to Turing machine programs.

C   *Criticisms of the Church-Turing thesis*

The Church-Turing thesis has its critics who suggest that it may well owe more to the "mathematics world-view",[73] prevalent at the inception of computer science as a discipline, rather than to any intellectual rigour. In particular, the definition of a Turing machine cannot account for interaction between the program and its environment, suggesting it "may no longer be fully appropriate to capture all

---

70 Barker-Plummer, above n 59.
71 Given the continued exponential growth in storage space evident throughout the whole history of computing, this does not seem an overly unrealistic assumption to make. For example, when the first hard drive was shipped by IBM in 1956, it was 5Mb in size. By 1980, IBM had introduced a 1000 Mb hard drive. By 2007, Seagate had built a 1,000,000 Mb hard drive. See Jon Wuebben, "The History of the Floppy and Hard Disk Drive" (20 March 2007) <http://www.patantconsult.com/articlesvault/Article/The-History-of-the-Floppy-and-Hard-Disk-Drive/11501> (11 September 2011).
72 Marvin L Minsky, *Computation: finite and infinite machines* (Prentice Hall, 1967), Chapter 11.
73 Dina Q. Goldin and Peter Wegner "The Church-Turing Thesis: Breaking the Myth" in *Proceedings of CiE 2005* 152 at 154–155.

features of present-day computing"[74] such as artificial intelligence, graphics, and the Internet. The difficulty with the Turing model, based as it is in mathematics, is in accounting for input after the program has started. As Tseytin puts it:

> An algorithm in the mathematical sense is completely self-contained and as soon as the data have [*sic*] been specified it needs no further information. In contrast to this, a realistic procedure (and, to some extent, a modern computer program) can draw information from the environment in a way that *need not be specified in advance*.[75]

Despite this, the Turing model can be extended rather than replaced in order to overcome such limitations.[76] Whilst these extended models might not be identical to the original, there is no evidence to suggest they are more or less powerful. So the Turing machine remains central in the theory of computer science. In any event, there is nothing about these extensions that necessarily breaks the equivalence between mathematics and software, since there is no reason why similar extensions could not be made to mathematics to encompass interactivity. In any event, an algorithm which collects all data beforehand is functionally equivalent to an algorithm which collects data interactively, so any difference does not alter the formal equivalence of mathematics and software.

Others have criticised the relevance of the Turing machine on the basis of industry practice. Dryja notes that the Turing machine fit neatly with computer software when it was primarily written by computer scientists, in a largely sequential form, almost a one-algorithm-per-program approach. Dryja claims that software development is now primarily the domain of engineers, centred around the top-down object-oriented programming approach described above. There is some truth in this contention, in that programs are not written as one long series of instructions anymore. But objects themselves merely "perform computations and save local state"[77] and an object-oriented program can be seen as a collection of interrelated Turing machines. It is, therefore, hard to accept that the Turing machine, and hence mathematics, is no longer relevant to an understanding of what software is. A collection of mathematical objects is no less mathematical in nature than its assembled whole.

What the above criticisms suggest is that the Turing machine model may not be complete account of what can be computed on a modern-day computer.

---

74 Jan van Leeuwen and Jirí Wiedermann, "The Turing Machine Paradigm in Contemporary Computing" in B. Enquist and W. Schmidt (eds), *Mathematics Unlimited – 2001 and Beyond* (Springer Verlag, 2000).

75 Gregory S. Tseytin "From Logicism to Proceduralism (An Autobiographical Account)" in *Algorithms in Modern Mathematics and Computer Science* (Springer Verlag: Berlin 1979) 390 at 395.

76 See for example the Persistent Turing Machine put forward in Goldin and Wegner, above n 73. Van Leeuwin and Wiederman offer an Interactive Turing Machine for the same purpose – see above n 74.

77 Mark Stefik and Daniel G. Bobrow, "Object-Oriented Programming: Themes and Variations," (1986) 6(4) *AI Magazine* 40 at 41.

Despite these criticisms, however, Turing machines remain fundamental in importance, since "they are widely used and have been widely adopted as the standard model in computability theory."[78] Thus, despite these criticisms, it is clear that the formal equivalence of software and mathematics, via the Turing machine, can be reliably claimed.

D   Structural similarities between mathematics and software

Further, there are structural similarities which suggest an equivalence between software and mathematics. In other words, there is a correspondence between structures found in software development and those used in mathematical activities.[79] If the activities engaged in by software developers and mathematicians are isomorphic, then the two are structurally equivalent. The starting point is the Curry-Howard isomorphism,[80] which asserts a correspondence between programming and mathematical logic. This isomorphism is a natural result of the formal equivalence of formal systems and the Turing machine noted above. Kondoh summarises it as follows:[81]

Table 1.4 The Curry-Howard isomorphism

Programming	Logic
Specification	Theory
Program	Proof

In a similar vein, the relationship between software and formal systems can be seen by comparing the definitions of programming languages and formal systems. First consider the following definition of a programming language:

> A set of *symbols*, understood by both sender and receiver, is combined according to a set of *rules*, its grammar or syntax. The semantics of the language defines how each grammatically correct sentence is to be interpreted.[82]

---

78 V.J. Rayward-Smith *A First Course in Computability* (Blackwell Scientific, 1986) at vii.
79 Hidetaka Kondoh, "What Is 'Mathematicalness' in Software Engineering?" in *Proceedings of Fundamental Approaches to Software Engineering 2000*, 163 at 170.
80 The full isomorphism is a combination of three seminal works by Curry and Howard. See Haskell Curry, "Functionality in Combinatory Logic" (1934) 20 *Proceedings of the National Academy of Sciences* 584; Haskell Curry and Robert Feys, *Combinatory Logic, Volume 1* (North-Holland, 1958) at paragraph 9E; William A. Howard, "The formulae-as-types notion of construction", in Jonathan P. Seldin and J. Roger Hindley (eds), *To H.B. Curry: Essays on Combinatory Logic, Lambda Calculus and Formalism*, (Academic Press, 1980) 479.
81 Kondoh, above n 79 at 170.
82 Alice E. Fischer and Frances S. Grodzinsky, *The Anatomy of Programming Languages 1* (1993). Cited in Gregory Stobbs, *Software Patents* (2nd ed, Aspen, 2000).

Then consider this definition of a formal system:

> Each formal system has a formal language composed of primitive *symbols* acted on by certain *rules* of formation (statements concerning the symbols, functions, and sentences allowable in the system) and developed by inference from a set of axioms. The system thus consists of any number of formulas built up through finite combinations of the primitive symbols—combinations that are formed from the axioms in accordance with the stated rules.[83]

The problem with the isomorphism at this level is that it assumes that programs can be "systematically derive[d] ... from their specifications",[84] an approach which meets the same criticisms Dryja levelled at the Turing machine, namely, that this isn't how programmers work. Kondoh is able to build on the Curry-Howard isomorphism, however, to demonstrate a wider equivalence between the modern art of software engineering and mathematics:[85]

Table 1.5 Kondoh's wider programming-mathematics isomorphism

Programming	Mathematics
Basic Control Structure (Repetitive Loop, Conditional, etc.)	Elementary Proof Step (Mathematical Induction, Case Analysis, etc.)
Idiom (Useful Combination of Control Structures)	Proof Technique (Conventional Technique for Fragmental Proof)
Abstract Data Type (Collection of Operations on Common Data)	Theory on a Mathematical Notation (Collection of Lemmata on a Mathematical Notion)
Design Pattern (Specific Combination of (Possible) Classes and Specific Use of Their Interdependencies)	Proof Tactics (Specific Combination of Subgoals (Lemmata) and Specific Use of Their Interdependencies)
Architectural Pattern (Specific Combination of Specification of Components)	Theory Strategy (Collection of Basic Definitions and the Main Theory)
Software System	Mathematical Theory (Structure Formed by Definitions, Theorems, and Proofs)

83 Encyclopaedia Britannica, "Formal system" *Encyclopaedia Britannica Online* <http://www.britannica.com/EBchecked/topic/213751/formal-system> (28 June 2011).

84 Kondoh, above n 79 at 170.

85 Kondoh, above n 79. See also Peter Suber, "Formal Systems and Machines: An Isomorphism" <http://www.earlham.edu/~peters/courses/logsys/machines.htm> (accessed 27 June 2011).

This similarity of structure between mathematics has led some computer scientists to apply mathematical proof techniques to computer programs in order to allow the verification of their results and confirm that they meet their specifications.[86] This sort of translation takes place in the opposite direction as well, with computers having become an indispensable tool by which mathematicians test theories through modelling[87] and the automation of formal proofs.[88] This further strengthens the notion that there is an intimate connection between the two disciplines.

4   Implications of the isomorphism for patent law

The underlying premise of patent law, whatever the justification,[89] is its claimed net-positive impact on innovation. Broadly speaking, innovation is "the implementation of a new or significantly improved product (good or service), process, new marketing method or a new organisational method in business practices, workplace organisation or external relations".[90] As such, it is asserted that deeper understanding of the impact of patent law on innovation is assisted by considering the nature of the activities undertaken by the relevant innovators, that is, mathematicians and programmers. Since the two domains share a formal and structural equivalence, this suggests that the effect of external factors such as the patent regime on innovative activities in these fields is likely to be equivalent.

It is for this reason that this book draws upon the treatment by patent law of mathematics, to justify a similar treatment for software. It is not asserted that mathematics and software are in all respects identical, only that they are closely related. The similarities and differences between the two are considered in much greater detail in Chapters 4 and 5.

86 See for example John McCarthy, "Towards a mathematical science of computation" in C.M. Popplewell (ed), *Information Processing 1962* (Holland Publishing Company, 1962) 21; <http://www-formal.stanford.edu/jmc/towards.html> (21 July 2008); C.A.R. Hoare "An axiomatic basis for computer programming" (1969) 12(10) *Communications of the ACM* 576.

87 For an introduction to the area see Wikipedia, "Computer simulation" <http://en.wikipedia.org/wiki/Computer_simulation> (4 November 2008).

88 For an introduction, see Wikipedia, "Interactive theorem proving" <http://en.wikipedia.org/wiki/Interactive_theorem_proving> (4 November 2008). For an ambitious project to build an automatically-verified wiki of mathematical knowledge, see Cameron Freer, "What is vdash?" <http://www.vdash.org/intro/> (11 September 2011).

89 See Introduction, at 7–8. Whilst the general justification discussed in this thesis is the social contract theory of patent law, this is not the only justification. Nonetheless, the notion that patents create or encourage innovation or at least the commercialisation of research, is a dominant claim of patent law.

90 OECD, Oslo Manual Guidelines for collecting and interpreting innovation data, (3rd edition, 2005). Cited Department of Innovation, Industry, Science and Research, "Introduction" *Australian Innovation System Report 2011*.

5   Conclusion

Two key themes have arisen from the discussion above. First, the history of software and the nature of software development demonstrate that three key considerations dictate the dimensions by which software must be studied. Successive iterations in the history of software have been delivered by layering **abstractions** upon abstractions, to the point where many programmers are no longer concerned in any meaningful fashion with the specific physical characteristics of the hardware their components might run on. Because the many layers of abstraction involve different levels of interaction with the physical hardware of the computer, consideration of a software component must consider its **context** in order to properly account for the physical constraints affecting the component's design, and the extent to which the component builds on the work of others. From the constant building new layers on previous layers developed a culture of **re-use**, which still pervades software development to this day, and without which the economics of software development would be unsustainable. Further, the **complexity** of software means that one application may be made up of a large number of interrelated but unique components, which for a large product may number into the millions. It will be shown in the next chapter that many of the problems which software has caused courts can ultimately be traced to these considerations.

Second, the equivalence of mathematics and software has been shown. The isomorphic nature of software and mathematics has been demonstrated at both a formal and structural level. Both the nature of software and the way in which programmers work are inherently mathematical in nature. In other words, either software developers are in some sense "doing mathematics" when they are writing software, or alternatively they are undertaking activities so similar in nature to mathematics that they give rise to the same sorts of policy considerations which apply to the activities of mathematicians. The patentability of software will be considered in later chapters through the lens of mathematics, and it will be demonstrated that the non-patentability of software follows from the non-patentability of mathematics.

2 Why software patents are a problem

The chance that a law will achieve its intended purpose improves when it is grounded in an accurate understanding of the phenomena it will regulate.[1]

This chapter will explain the software patent problem by defining what a software patent is and why it is that software patents are such a difficult fit within the patent regime, tracing these difficulties to the key considerations identified in Chapter 1.[2] First, it was noted that software has an abstract, intangible nature, as opposed to the hardware on which it runs. As such, it is important to consider context, namely, the type of software component being developed and how many layers of abstraction sit between it and the hardware. Second, it was noted that reuse is critical in software development, since new software innovations almost inevitably build on what has come before. Finally, software products are highly complex, involving millions of discrete components. The way in which these factors combine to produce the software patent problem is set out below.

The chapter will also document the way in which the courts of the US, the EU, and Australia have attempted to deal with these problems, and describe why it is that the software patent issue should be dealt with as a patentable subject matter issue rather than as an issue of novelty or non-obviousness.

It will be shown why the current approaches are insufficient, and to introduce the notion that a fresh approach to the software patent problem is needed.

1 What is a software patent?

A threshold issue is that there is no agreed definition of what a software patent is. A broad definition of software might include anything that is not hardware, hardware being the "physical component of a computer system",[3] or the "part[s] of a computer that is fixed and cannot be altered without replacement or physical

---

1 Pamela Samuelson, "Foreword: The Digital Content Symposium" (1997) 12(1) *Berkeley Technology Law Journal*. <http://www.law.berkeley.edu/journals/btlj/articles/vol12/Samuelson/html/reader.html> (25 February 2008).
2 See Chapter 1, at 20–23.
3 "Hardware" in M.J. Clugston et al. (eds), *The New Penguin Dictionary of Science* (Penguin, 2004).

modification".[4] Following on from this definition, then, a software patent would be a patent for a computer-based invention which can be altered without replacement or physical modification. An alternative definition is that offered by Allison and Lemley, who define a software patent as "[a]n invention that is completely embodied in software, even if the claims of the patent refer to a system or article of manufacture".[5] Such a definition excludes inventions which include software as a component of a physical device.

The difficulty with definitions such as these is twofold. First, they underestimate the extent to which the physical elements of a computer-based innovation can directly influence related software components. A good example is firmware, discussed in Chapter 1, which is embedded in a hardware device and is functionally limited to interaction with that hardware, but can be modified via a "flash upgrade" and thus has the modifiability of software.

Second, by focusing on 'pure' software patents, one runs the risk of failing to consider patents drafted to appear as physical devices in which software is merely a component, but which in substance are directed to software.[6]

For present purposes, then, it is important consider as wide a range of software and software-related innovations as possible. This book therefore adopts a broader view of what amounts to a software patent, namely:

> [A] logic algorithm for processing data that is implemented via stored instructions; that is, the logic is not "hard-wired." These instructions could reside on a disk or other storage medium or they could be stored in "firmware," that is, a read-only memory, as is typical of embedded software.[7]

2   Why is the patenting of software problematic?

2.1   A history of uncertainty

The different historical origins of the modern patent regimes in Australia, the US, and UK have resulted in very different approaches to the issue of patentable subject matter. Although Australia and the UK have a long shared history, the adoption of the *European Patent Convention*[8] by the UK in 1977 forever broke that link.[9] Whereas the Australian position continues to define patentable subject matter by reference to the definition in the *Statute of Monopolies 1623*

4 "Hardware" *Wiktionary* <en.wiktionary.org/wiki/hardware> (6 November 2008).
5 John R. Allison and Mark A. Lemley, "Who's Patenting What? An Empirical Exploration of Patent Prosecution" (2000) 58 *Vanderbilt Law Review* 2099 at 2110.
6 See the discussion below in Section 2.2, where claims have been directed to apparatus executing the software, or to a physical carrier containing the executable software. See also James E. Bessen and Robert M. Hunt, "An Empirical Look at Software Patents" (2007) 16(1) *Journal of Economics & Management Strategy* 157 at 163.
7 Bessen and Hunt, above n 6 at 163.
8 *Convention on the Grant of European Patents*, opened for signature 5 October 1973, 13 ILM 268 (entered into force 7 October 1973) ("EPC").
9 At least in the patent law context.

(UK), and the notion therein of a manner of manufacture, the UK position is now defined by the express exclusions in Article 52 of the EPC, and their interpretation by the European Patent Office ("EPO") as resolving to a single requirement of technical character. The US approach, although likely influenced by the *Statute of Monopolies*, derives uniquely from the power afforded to Congress in the US Constitution to pass laws to "promote the Progress of Science and useful Arts",[10] and is determined by the four statutory classes of patentable subject matter defined in 35 USC §101, namely, processes, machines, manufactures, and compositions of matter.

Despite these differences in approach, all three jurisdictions have struggled with drawing a stable boundary line. The global dominance of the United States in computer hardware and software development might suggest that after 40 years of litigation and the associated development of jurisprudence, the issue might be resolved in that jurisdiction at least. Until relatively recently, it seemed as though the issue had been resolved in favour of the unfettered patenting of software.

However, recent Supreme Court decisions have seen the pendulum swing markedly away from the patentability of software. Notably, in 2014 in *Alice Corp.*,[11] the Supreme Court drew a distinction between patent ineligible abstract ideas and patent-eligible applications of those ideas. The practical effect of *Alice Corp.* has been that in many if not most cases, subject matter thought previously to be patentable has been rejected as directed to the former.

As a counterpoint we have the United Kingdom, whose initially favourable view of the patentability of software was turned on its head by the passage of the EPC, which expressly excluded computer programs "as such" from patentable subject matter. The subsequent watering-down of the exclusion, especially at the EPO, however, highlights the difficulty in defining the scope of patentable subject matter, either in positive or negative terms.

Australian patent law decisions have, at least broadly speaking, tracked developments in the US. As such, the most recent decisions in the Full Federal Court[12] have questioned whether claimed inventions are, "as a matter of substance, not form",[13] directed to a non-patentable abstract idea or are instead "technical" in nature.[14]

It is suggested that the uncertainty which flows from this constant movement of the boundaries of patentable subject matter derives from the approach of courts and tribunals to the characterisation of software patent claims and assumptions therein about the nature of software. It will be shown how these varied approaches have contributed to an essentially identical software patent problem which manifests itself in theoretical problems for patent law and

---

10 United States Constitution, Article One, Section 8, Clause 8.
11 *Alice Corporation Pty Ltd v CLS Bank International* 134 S. Ct. 2347 (2014) (*"Alice Corp"*).
12 *Research Affiliates LLC v Commissioner of Patents* (2014) 227 FCR 378 (*"Research Affiliates"*); 109 IPR 364, and *Commissioner of Patents v RPL Central Pty Ltd* (2015) 238 FCR 27; 115 IPR 461 (*"RPL Central"*).
13 *RPL Central* at [98].
14 *RPL Central* at [97], [99].

practical problems for the software industry. These problems will be considered by reference to the four key elements of the nature of software developed in the last chapter, namely, abstraction, context, complexity, and reuse.

2.2   Abstraction

As was discussed in Chapter 1, the history of software is one of increasing layers of abstraction between software and the hardware on which it runs. All but firmware and the lowest levels of the operating system are implemented largely independently of the physical machine on which they run.

Thus there is a tension with one of the classic doctrines of patent law, namely that patents are supposed to protect the execution or application of an idea in the form of an invention, not the idea itself. As noted above,[15] this is the dominant approach in the US and Australia at the present time, although this approach is nothing new.[16] When an invention has a physical manifestation, it is easy to distinguish the invention from the idea behind it. However, patent law long outgrew such straightforward physical inventions. The move to ever more abstract subject matter began with patents for improvements to these machines and devices, expanded then through methods of effecting such improvements and new uses of existing machines and substances,[17] before arriving at the information-based innovations of the present age.

In *Hickton's* case,[18] the Court suggested that an abstract idea becomes patentable subject matter when the claims disclose "a way of carrying [the idea] out".[19] This is particularly problematic for software. It will be recalled from Chapter 1 that the software development process involves a gradual progression from the idea behind a program towards code actually implementing it. Knowing at which point the non-patentable idea has been expressed in sufficiently concrete terms so as to amount to an invention is a difficult task. What was already "a question of degree"[20] becomes a task of splitting hairs, encouraging the abandonment of the distinction altogether.[21]

---

15 See Section 2.1 above.

16 A non-patentable or "abstract" idea was the basis for exclusion in *O'Reilly v Morse* 56 US 62 (1852), and has also been considered by reference to the terms "principle": *Boulton v Bull* (1795) 126 ER 651 ("*Boulton v Bull*"); *Le Roy v Tatham* 55 US 156 (1852), and "abstract notion": *Boulton*.

17 See the discussion of the non-patentability of analogous uses in in *National Research Development Corporation v Commissioner of Patents* (1959) 102 CLR 252 ("NRDC").

18 *Hickton's Patent Syndicate v Patents and Machine Improvements Co. Ltd* (1909) 26 RPC 339 ("*Hickton's*").

19 *Hickton's* at 348 per Fletcher-Moulton LJ.

20 *Harwood v Great Northern Railway* (1865) 11 ER 1488 at 1499.

21 This is arguably what happened in more recent European Patent Office jurisprudence in which the computer program exclusion has been reduced to a mere form provision. An overview of this "any hardware" approach is set out in the discussion of the case *Aerotel Ltd v Telco Holdings Ltd and in the matter of Macrossan's Application* [2006] EWCA Civ 1371; [2007] RPC 7 ("*Aerotel*"). See also section 4 below.

The same problem arises when drawing a line between a discovery and its practical application,[22] where the suggested application is itself abstract in nature.[23] Such patents, due to their breadth, are prone to awarding control over a wide range of independently developed technologies – a windfall for the patent holder at the expense of other innovators. As a result, the general public suffer from the slowing of "the onward march of science".[24]

If the computer industry continues the trend towards abstractness set out in Chapter 1,[25] these problems will only be exacerbated. Further abstractions will continue to be layered on top of each other, bringing programming languages ever closer to natural language, and making it harder to discern the difference between programming and user interaction.[26]

This is not to say that courts are unaware of the dangers of over-protection. A range of responses to prevent the award of patents "so abstract and sweeping as to cover both known and unknown uses"[27] have been developed, with the US Supreme Court decision in *Alice Corp.* being the most recent example. It will now be shown how the approaches adopted have failed to provide an effective safeguard.

22 See for example *Gottschalk v Benson* 409 US 63 (1973) (*"Gottschalk v Benson"*) where it was held that allowing the claim in the absence of a specific practical application would amount to a patent on an idea; see also *Parker v Flook* 437 U.S. 584 (1978) (*"Flook"*) where it was held that the required application must be more than insignificant post-solution activity; *Bilski v Kappos* 130 S. Ct. 3218 (2010) (*"Bilski"*) applied these approaches in holding the subject claims to be a non-patentable abstract idea; and most recently *Alice Corp*, Cf. *Burroughs Corporation's Application* [1973] FSR 439 (*"Burroughs"*) at 449 where a "naked conception" or "bare method or idea" was contrasted with "practical embodiment" of software "enabling that method to be realised in practice". Similarly in *Genentech Inc's Application (Human Growth Hormone)* [1989] RPC 147 at 240 it was held that "the practical application of a discovery" was patentable subject matter since such a claim does not relate to the discovery "as such". In the Australian context, see *NRDC* at 264, where the court characterised a non-patentable discovery as "abstract information without any suggestion of a practical application".
23 That this context is important consideration is further developed below.
24 *Gottschalk v Benson* at 68.
25 This seems probable, at least in the short to medium term, absent some disruptive new methodology superseding the existing approach.
26 Some significant headway has already been made in such areas as speech recognition and in building systems to parse written language. But as anyone who has used interactive voice recognition menus from their phone knows, the current state of the art is far from perfect: see Robert Fortner, "Rest in Peas: The Unrecognized Death of Speech Recognition" <http://robertfortner.posterous.com/the-unrecognized-death-of-speech-recognition> (2 August 2010); Peter Bradley, "Turing Machines and Natural Language" *Consortium on Cognitive Science Instruction*, 2002 <http://www.mind.ilstu.edu/curriculum/turing_machines/turing_machines_and_language.php> (2 August 2010). Although they may not yet be perfect, voice interactions with devices are fast approaching error rates comparable to humans: see W. Xiong et al., "Achieving Human Parity in Conversational Speech Recognition," Microsoft Technical Report MSR-TR-2016-71, revised February 2017.
27 *Gottschalk v Benson* at 68.

A   Pre-emption/breadth

In the UK, the House of Lords in *British United*[28] suggested a broad claim was
"not proper subject matter"[29] because a claim ought to "inform others of the
limits of such protection".[30] A similar view was expressed in *RCA Photophone*[31]
where this requirement was held to be the "consideration" for the monopoly.[32]
It was also noted that an overbroad claim also creates "a public nuisance".[33] A
similar limitation arises from the judgment of Parker J in *British United*, where
his Honour asserted that no authority existed for the proposition that the disco-
verer of a new principle "could have protected himself against all means of
solving [a] problem".[34] This line of reasoning also appears to inform the reasoning in subsequent case law,[35] although was subjected to a complicating and
largely unhelpful caveat in *David Kahn*[36] that such subject matter might not
in fact be excluded where "invention lies in identification of the problem".[37]

Along similar lines, the US Supreme Court in *Gottschalk v Benson* limited
claims involving mathematical algorithms where such claims, due to the
absence of a practical application, would "wholly pre-empt the mathematical
formula and in practical effect would be a patent on the algorithm itself".[38]
Although subsequent interpretation of this requirement lead to a narrow requirement that such claims first be directed to a mathematical formula or its equivalent,[39] the US Supreme Court in *Bilski* in 2010 brought the pre-emption issue
back to the fore.[40]

By only looking to the practical effect of a particular patent to determine
whether it passes muster, however, means no overarching principles can be developed. A case-by-case approach denies the potential usefulness of considering
exclusions at the categorical level, and engaging with underlying policy

28 *British United Shoe Machinery Co Ltd v Simon Collier Ltd* (1909) 26 RPC 21 (*"British United"*).
29 *British United* at 50.
30 *British United* at 50.
31 *RCA Photophone Ltd v Gaumont-British Picture Corporation* (1936) 53 RPC 167 at 186–187 (*"RCA Photophone"*).
32 This is of course informed by the view of patent law as a social contract. See the discussion of the social contract in *RCA Photophone*; and the discussion in *Liardet v Johnson* (1778) 1 Carp Pat Cas 35 (NP).
33 *RCA Photophone* at 194.
34 *British United* at 50.
35 See for example *Gale's Patent Application* [1991] RPC 305 (*"Gale"*) at 327–28 where it was held that the claimed invention was not patentable, *inter alia*, because it did not "solve a 'technical' problem lying within the computer".
36 *David Kahn Inc v Conway Stewart* [1974] RPC 279 (*"David Kahn"*).
37 *David Kahn* at 319–320.
38 *Gottschalk v Benson* at 72. This concern also informs the rejection of insignificant post-solution activity in *Flook*. See also *Bilski* at 3230 per Kennedy J.
39 See *In re Freeman* 573 F.2d 1243 (1978) at 1246.
40 See *Bilski* at 3239 per Kennedy J: "[T]he Court resolves this case narrowly on the basis of this Court's decisions in *Gottschalk v Benson*, *Flook* and *Diehr*, which show that petitioners' claims are not patentable".

considerations which might then reduce uncertainty and provide future guidance in the area. It also fails to acknowledge that categorical exclusions have been part of patent law since at least the passage of the *Statute of Monopolies*.[41]

Further, the starting point in this analysis is a general presumption that patents should be available, with exceptions for particular instances. The policy choices which inform such a presumption are not considered. Identification of breadth and pre-emption as the source of the problem does not resolve the issue, as the harm these doctrines seek to overcome is never specified. Why is it that a claim to an algorithm *should* be considered non-patentable? Put simply, the case law just discussed "never provides a satisfying account of what constitutes [an overbroad claim]"[42] and "essentially asserts [a] conclusion".[43]

B   Mental steps doctrine

Another approach adopted in the US was to disallow patentability under the "mental steps doctrine".[44] This doctrine recognised that claims directed to highly abstract subject matter as "intangible, illusory and non-material things"[45] run an appreciable risk of infringement by human operators carrying out the claimed process. In *Halliburton v Walker*,[46] the Court ruled ineligible a method of determining the location of obstruction in an oil well. The Court emphasised the dependence on descriptive words in the claims, such as "'determining,' 'registering,' 'counting,' 'observing,' 'measuring,' 'comparing,' 'recording,' 'computing'",[47] holding that such "mental steps, even if novel, are not patentable".[48]

41 See for example Coke's explanation of the phrase "contrary to law" in his *Institutes of Law*, as excluding improvements to existing manufactures, an exclusionary ground which Coke traces back to *Bircot's case*, decided in 1572.

42 *Bilski* at 3236 per Stevens J.

43 *Bilski* at 3236 per Stevens J.

44 This is not to say that its relevance is limited to that jurisdiction. The EPC excludes the patentability of "schemes, rules and methods for performing mental acts ...and programs for computers" which might suggest a similar relationship between the two. The examiner in *Fujitsu Limited's Application* [1996] RPC 511 ("*Fujitsu (No 1)*") rejected the claimed invention as falling within one or both exclusions. Laddie J, in reviewing the decision of the examiner, held that the computer program exclusion had been avoided, but the claims were directed to a mental act, since they involved "a significant level of abstraction and generality": at 532. Note also that the relevance of claims reading on mental steps was acknowledged in the pre-EPC case of *IBM's Application* [1980] FSR 564 at 567 where it was said "such an operation could in theory be done without the need for any automatic aids but in practice needs to be automatically computed." Given the strong endorsement of this case in Australian jurisprudence, this arguably represents the position in Australia. This position is further fortified by the exclusion for schemes, rules and plans, which find expression in the same section of Article 52 EPC as the mental steps doctrine, suggesting a close relationship between these areas.

45 *Greenewalt v Stanley Co of America* 12 USPQ 122 (1931) at 123.

46 *Halliburton Oil Well Cementing Co v Walker* 146 F.2d 817 (1944) ("*Halliburton*").

47 *Halliburton* at 821.

48 *Halliburton Oil Well Cementing Co v Walker* 146 F.2d 817 (1944) at 821.

The application of the doctrine becomes more complicated where the steps claimed involve a combination of "positive and physical steps as well as so-called mental steps",[49] since the involvement of human operators had previously been considered unproblematic. Cases such as *Abrams* held that the reasons for the exception were "self-evident".[50] The lack of a solid theoretical underpinning was to prove the doctrine's downfall.[51]

In the software context, the doctrine faces an intuitive hurdle, namely that claims directed to a machine could be reasonably interpreted as reading upon human mental activity. The notion that a human mind operated in the same way as a computer was explicitly rejected in *Bernhart*[52] but held to be beyond a "reasonable interpretation of the claims" in the decision of the Court of Customs and Patent Appeals (CCPA) in *Benson*.[53] In that case, the fact that a computer could carry out the process "without human intervention"[54] was sufficient to avoid the exclusion. In this sense the approach of the CCPA is similar to the EPO's interpretation of the computer program exclusion, wherein the execution of the program on computer hardware is sufficient to avoid the exclusion.

Assumptions about the relationship between computer and mind and the relationship between patent law and the First Amendment naturally arise here,[55] but

---

49 *In re Abrams* 89 USPQ (BNA) 266 (1951) (*"Abrams"*).
50 *Abrams* at 269.
51 One possible foundation was put forward in *In re Prater* 415 F.2d 1378 (1968) (*"Prater (No 1)"*), namely inconsistency with a right of freedom of speech, but it was never explored in any detail. The possible relevance of this foundation is explored later in this thesis.
52 *In re Bernhart* 417 F.2d 1395 (1969) (*"Bernhart"*) at 1401.
53 *In re Benson* 441 F.2d 682 (1971) at 687.
54 *Prater (No 1)*; *In re Prater* 415 F.2d 1393 (1969) (*"Prater (No 2)"*) at 1403. This notion underpins the later developments in *Bernhart* and *In re Musgrave* 431 F.2d 882 (1970) (*"Musgrave"*).
55 The Patent Office in *Prater (No 2)*, sought a rehearing of the earlier rejection of the mental steps doctrine in *Prater (No 1)*, on the basis that the Constitutional issues had not been dealt with. In particular, it was said that the subject patent would "confer upon a patentee the right to exclude others from thinking in a certain manner": *Prater (No 2)* at 1401 (footnote 20). It was also contended that the patent fell foul of the Ninth and Tenth Amendments. The Ninth Amendment states that "[t]he enumeration in the Constitution, of certain rights, shall not be construed to deny or disparage others retained by the people". As such it operates not as a positive source of rights, but negatively, to prevent the enumerated rights in the Constitution being used as the basis of the denial of non-enumerated rights. As such the amendment would arguably prevent the enumerated freedom of speech from being interpreted in such a way as to deny freedom of thought, on which freedom of speech depends (the relationship between freedom of thought and freedom of speech/expression is considered further in Chapter 4). The Tenth Amendment, known as the reserve powers clause states that "[t]he powers not delegated to the United States by the Constitution, nor prohibited by it to the States, are reserved to the States respectively, or to the people". As such, the argument would seem to be that the people could not have been taken to have delegated the power to regulate their thoughts to Congress. Similar

have never been given direct consideration.[56] In *Musgrave* the doctrine was dismissed in favour of a requirement that the claimed invention "be in the technological arts".[57] The Supreme Court in *Gottschalk v Benson* had an opportunity to address the mental steps doctrine but redirected attention to the patentability of mathematical algorithms.

The relationship between computers and mental processes is both complex and important.[58] It is therefore unfortunate that this doctrine was abandoned,[59] as it forced the consideration of this interplay directly. Such consideration might have directly addressed the admittedly difficult issues of the relationship between the computer and the human mind[60] and whether it is appropriate that patent law by "conferring a broad property right in cognitive process steps appropriates publicly necessary information to private use".[61]

It may be that the diversion of attention to the patentability of algorithms really only amounts to a change in nomenclature.[62] But this just introduces a similarly troublesome distinction between mathematics and software development which encourages speculation about what is mathematical and what is not, and about the differences between pure and applied mathematics.

---

support for freedom of thought may perhaps be sourced in the "penumbra" and "emanations" of the right of free speech: see *Griswold v Connecticut* 381 US 479 (1965) per Douglas J; or as an incident of the "concept of personal 'liberty' embodied in the Fourteenth Amendment's Due Process Clause": *Roe v Wade* 410 US 113 (1973) at 129. This topic will be explored in further detail in Chapter 4.

56 On the rehearing, the Court determined that it did not need to engage in such analysis, as in any event the claimed invention fell outside the ambit of the mental steps doctrine: *Prater (No 2)* at 1403.

57 *Musgrave* at 893. The interrelationship between this doctrine and the technological or useful arts limitation will be advanced later in this book.

58 This interplay is explored in great detail later in this book.

59 Samuelson notes that with the exception of *In re Meyer* 688 F.2d 789 (1982) and *In re Sarkar* 588 F.2d 1330 (1978) ("*Sarkar*") "the 'mental process' rationale for rejecting program-related inventions rarely appears in the case law.": Pamela Samuelson, "Benson Revisited: The Case Against Patent Protection for Algorithms and Other Computer Program-Related Inventions," (1990) 39 *Emory Law Journal* 1025 at 1043, note 59.

60 "That a human being is a kind of machine and that the human mind is (or is like) a digital computer are among the central theories of cognitive science and of the artificial intelligence disciplines.": Samuelson, above n 59 at 1044, note 60. See also *ibid* at 1060, note 122: "[T]he phrases 'mental process' and 'mental steps,' while admittedly awkward and not entirely apt for describing some important patentable subject matter problems, may be closer to the heart of the problem with software patentability than the term 'algorithm,' which in the first post-Benson case law comes to dominate the debate." See also Allen Newell, "Response: The Models Are Broken, The Models Are Broken," (1986) 47 *University of Pittsburgh Law Review* 1023 at 1025.

61 Dan L. Burk, "Patenting speech" (2000) 79 *Texas Law Review* 99 at 140.

62 "Sequences of mental steps and algorithms are the same thing. Any attempt in the law to make distinctions that depend upon contrasting mental steps versus algorithms is doomed to eventual confusion.": Newell, above n 60 at 1025.

C   Intellectual information exclusion

A similar approach developed in UK law, which held mere intellectual information to be non-patentable subject matter.[63] This same exclusion has been
adopted in Australia, although rejections are typically referred to other exclusionary grounds, such as the lack of a practical application,[64] or as falling within the
fine arts.[65] Such an exclusionary basis also exists in the US, where it is known as
the printed matter doctrine.[66] Although the name suggests a different focus, the
substantive issue addressed is the same.

On a theoretical level, this exclusion runs into trouble because "[p]atents are
essentially about information as to what to make or do."[67] Thus the subject
matter of patents is intangible. Although patent claims may describe the features
and interrelationship of physical entities, these descriptions are themselves only
representative. In other words, the determination of subject matter issues is conducted at an abstract level. Is it any wonder then that abstract claims, bearing a
mere representational link to a physical entity, might seem sufficient?

---

63 See for example *Re Cooper's Application* (1901) 19 RPC 53 (AG) (*"Cooper's Application"*).
Similarly in *Fishburn's Application* (1940) 57 RPC 245 the arrangement of information on
both halves of a ticket so as to allow it to be torn in half without losing the requisite information was held to be patentable as it served a mechanical purpose). In contrast, in *Pitman's
Application* [1969] RPC 646, a method of teaching the pronunciation of language by visual
means was held to serve a functional purpose and thus constitute patentable subject matter.
In the UK, this exclusion is now statutorily enacted, as required by the European Patent
Convention. In *Fujitsu (No 1)* both Hearing Officer Haselden and Laddie J on appeal
rejected program as concerned with information content/intellectual process, although
the Court of Appeal (*Fujitsu Ltd's Application* [1997] RPC 608 (*"Fujitsu (No 2)"*))
looked instead to the absence of a technical contribution. It may be that these are but
two sides of the same coin.
64 *Grant v Commissioner of Patents* (2006) 154 FCR 62 (*"Grant"*) at 65–66: "It has long been
accepted that 'intellectual information' ... [is] not patentable." See *NRDC* at 264, where
the court stated that a non-patentable 'discovery' would lie where there was only "abstract
information without any suggestion of a practical application".
65 The distinction between useful and fine arts distinction was set out in *NRDC* at 275. That
intellectual information falls within the fine arts was explicitly stated in *Virginia-Carolina
Chemical Corp's Application* [1958] RPC 35. See also *CCOM Pty Ltd v Jiejing Pty Ltd*
(1993) 27 IPR 577 at 594, cited in *CCOM Pty Ltd v Jiejing Pty Ltd* (1994) 51 FCR
260 (*"CCOM"*) at 286, where the trial judge noted that "[t]he formulation of ...criteria
and their use as rules to organise and process data... are the product of human intellectual
activity lying in the fine arts and not the useful arts."
66 "The mere arrangement of printed matter on a sheet or sheets of paper ...does not constitute 'any new and useful art, machine, manufacture or composition of matter' [as required
by the US Patents Act]": *In re Russell* (1931) 48 F.2d 1395. As with UK cases discussed in
above, where a claimed invention demonstrates a functional purpose, also phrased as
'mechanical inventiveness', it will not fall foul of this exclusion. See for example *Flood v
Coe* 31 F Supp 348 (1940) at 349 where the "unique relationship between the physical
structure and the printed matter" on claims for a price tag meant that it amounted to
"more than an arrangement of printed matter on a piece of paper".
67 *Aerotel* at [32].

This issue is even more pronounced where the subject matter is the intellectual process of designing and implementing software ideas, which take as their input an abstract idea and descend through various levels of abstraction until arriving at an intangible output.[68] It is here, where the distinctions are the slightest, that the most resolute adherence to them is required.

Theoretical problems aside, applying this exclusion means trying to distinguish the content from the presentation of information.[69] Characterisation of an invention becomes a task of great importance. An example is *Cooper's Application*, where the former was held non-patentable *inter alia* because it belongs to the realm of copyright law.[70]

Since the time of *Coopers' Application*, however, the exclusion has been interpreted narrowly, dispensing with the requirement of a physical product in favour of a "human act of inventiveness and that such act be the source of the subject matter's demonstrated practical utility".[71]

In the context of computer programs, this exclusion devolves to attempting to delineate between functional and informational aspects; distinguishing code from data. An example is *Badger's Application*[72] where a distinction was drawn between the "preparation, tabulation and codification of data" which was held to be "conceptual in character, lacking in … concrete actuality which differentiates substance from notion",[73] whereas the "conditioning of the computer to permit computation"[74] (that is, the code) was found to be appropriate subject matter. Similarly, in *Slee & Harris* the term "data" was held to be equivalent to intellectual information.[75]

Computer scientists have long understood, however, the difficulty in distinguishing between code and data. A simple example serves to illustrate the point. Building on the loyalty program example from the *Catuity* case,[76] one might assume that the number of points awarded per dollar spent is different in each store. This information might be represented as data in the following table:

---

68 See Chapter 1, Section 3. See also Gary Dukarich, "Patentability of Dedicated information Processors and Infringement Protection of Inventions that Use Them" (1989) 29 *Jurimetrics Journal* 135 at 140–143.

69 In *Nelson's Application* [1980] RPC 173 at 176 the question was said to be "whether a claim …is a claim to a manner of new manufacture or whether it is really directed to monopolising something which is of a purely intellectual nature".

70 The relationship between copyright and patent law was explored (albeit from the copyright perspective) in *Baker v Selden* 101 US 99 (1879) in which it was held that the use of "useful art" (a system of book keeping) explained in a book was not copyright infringement, and could only be protected through patent law. See also Burk, above n 61 at 142–144.

71 Justine Pila, "Inherent patentability in Anglo-Australian law: a history," (2003) 14 *Australian Intellectual Property Journal* 109 at 162–163.

72 *Badger Co Inc's Application* [1969] FSR 474 ("*Badger's Application*").

73 *Badger's Application* at 476.

74 *Badger's Application* at 476.

75 *Slee & Harris' Application* [1966] FSR 51 ("*Slee & Harris*") at 53.

76 *Welcome Real-Time SA v Catuity Inc and Ors* (2001) 113 FCR 110 ("*Catuity*").

Table 2.1 Frequent flyer points look-up table

Store	Points per dollar
Grocery retailer	2
Hardware chain	3
Airline	1

The code to look up the points per dollar, given a particular store name may then look as follows:

```
def points_per_dollar(store_name):
    return lookup_table(store_name, column=2)
```

In these circumstances, the distinction between code and data is easily seen. The data is what is in the table, and the code defines the process by which that data is extracted. However, the same functionality might also be expressed entirely as code (that is, without using a table):[77]

```
def points_per_dollar(store_name):
    if store_name == Grocery retailer:
        return 2
    else if store_name == Hardware chain:
        return 3
    else:
        return 1.5
```

Here, where the data is embedded in the code, the distinction between code and data is less clear. Further enhancements to the points scheme can be imagined which might blur the distinction even more:

1   where the hardware chain offers double points from 1 January to 14 February each year; and
2   where there is interlinking between stores, for example where a purchase at both the grocery retailer and airline in the same months gets a 500-point bonus.

Using the second approach for anything other than a small dataset is bad software design, but the example illustrates the point nonetheless. The two approaches are functionally equivalent, and the claims which cover them would

---

77 Programmers seeing this code will no doubt it describe it as "hacky" or "ugly", but it does illustrate the point. The aesthetics of code is considered in more detail in Chapter 5.

be identical. So any attempt to draw generalised distinctions between code and data fails in this case.

In case the author is accused of manufacturing an edge-case to fit his argument, consider the following series of numbers:

0, 1, 1, 2, 3, 5, 8, 13, 21, ...

This is the Fibonacci series, where each number is equal to the sum of the two numbers preceding it in the series. The Fibonacci series appears in real-world phenomena, and is useful in creating computer-based models of them. One way of storing this series would be as a look-up table, just like the loyalty program table above:

Table 2.2 Fibonacci series look-up table

Fibonacci number
0
1
1
2
3
. . .

This may be a suitable approach where the number of elements of the series to be stored is small. For a large series, it might be better to calculate each row on demand, like this:[78]

```
define fib(n):
    if n == 0:
        return 0
    elif n == 1:
            return 1
    else:
        return fib(n-1) + fib(n-2)
```

Complex data is the norm in software development, since "[p]rograms are typically designed to model complex phenomena, and more often than not one

---

78 It wouldn't be written in this form in practice, as the recursive nature of the definition means that to calculate the 100th element of the series in this form "requires 708449696 358523830149 function calls, which would take longer than the age of the universe to execute even though the stack depth would never exceed 100": "Fibonacci numbers (Python)" <http://en.literateprograms.org/Fibonacci_numbers_(Python)> (24 August 2010).

must construct computational objects that have several parts in order to model real-world phenomena that have several aspects."[79]

This complexity itself gives rise to more complex interrelationships.[80] Code and data may also be entirely interchangeable labels depending on context. For example, from the perspective of the programmer writing a compiler or interpreter for a particular programming language, programs written in that language are just data to be fed into the compiler. Similarly, executable software, or object code, is data which the compiler or interpreter produces.

In fact, this implicit interchangeability lies at heart of theoretical foundation of modern computing. It is inherent in the definition of a Turing machine that such a machine is powerful enough to treat its own code as data.[81]

Australian cases directed towards computer-related inventions have glossed over these intellectual aspects of computer software in favour of a focus on the end result of its execution.[82] The problem with focusing on the end result is that it obfuscates the distinction between hardware and software, by looking to the physical effect of execution rather than considering software itself as a standalone entity. In other words, there has been a focus on the ends rather than the means.[83]

Doctrines of pre-emption, mental steps and intellectual information touch upon the right issues. Where they have failed is in a lack of direct engagement with the underlying issues. Instead they jump to conclusions, rely on illusory distinctions, or skate off into tangential inquiries. Meeting the challenge of a theoretical underpinning to abstract subject matter is something to which the latter part of this book is directed.

---

79 Harold Abelson and Gerald Jay Sussman, "Chapter 2: Building Abstractions with Data" in *Structure and Interpretation of Computer Programs* (2nd ed, 1996) at 13.

80 At the simplest end of the complexity scale would be whole numbers. These can become more complex by their combination into compound data types such as lists, series, and trees (or hierarchies). A level of abstraction above this is to make numbers representative of other things, such as characters, pixels, colours, and so on. Further complexity is possible when compound data types are capable of containing arbitrary data types, wherein the code processing the data would be the same for a list of numbers, a list of Greek letters, or even a list of lists. See Abelson and Sussman, above n 79.

81 See Piers Cawley, "Code is data, and it always has been" on *Just a Summary* <http://www.bofh.org.uk/2008/04/07/code-is-data-and-it-always-has-been> (1 September 2010). Cawley draws this inference from Turing's description of the halting problem, which assumes that it is possible for a Turing machine to treat a program written for it as data which can be fed into a second program. The second program in the halting problem context, is a program which is to detect whether the first program will ever reach a conclusion. Turing's proof establishes that such a program cannot be written, but the ability to treat a program as data for a second program is a necessary precondition of the proof.

82 See *CCOM* at 295 where the Full Federal Court interpreted *NRDC* as requiring only "a mode or manner of achieving an end result which is an artificially created state of affairs of utility in the field of economic endeavour". In *Grant* at 70, the Court added a further requirement that a "concrete, tangible, physical, or observable effect" be produced.

83 *RCA Photophone Ltd v Gaumont-British Picture Corporation* (1936) 53 RPC 167 ("*RCA Photophone*") at 191.

Given the potential problems which highly abstract subject matter creates, it becomes clear that considering the context of the claimed invention, determining the level of abstraction and relationship with the physical aspects of the device, is of particular importance.

The term context could perhaps be replaced by physicality, in that the physicality of a claimed invention is an important consideration in determining the nature of what is said to have been invented. Similarly, the notion of context could be subsumed within the discussion of abstraction above, as the absence of physicality might be taken to be confirmation of an alleged invention's abstract nature. But looking at the context in which a claimed invention is to operate is an inquiry preliminary to the determination of whether an invention is abstract or sufficiently limited by physical constraints, since it is not until one knows what advance is claimed that one can determine why it is or is not patentable subject matter.

Identifying context requires identification of the contribution or advance which the invention entails. This is said to require an examiner to look beyond the form of the claims to their "substance".[84]

Critics of a substantive analysis propound two concerns. The first is that such issues intrude upon other indicia of patentability, namely novelty and inventive step. This interrelationship is not surprising, given that these indicia originally co-existed with inherent patentability as a single consideration. It was only later in the history of patent law that they gained any independent existence. In fact, for this reason it has been asserted that the software patent problem is in fact an inventive step issue rather than one to be dealt with under the patentable subject matter penumbra. That claim is dealt with in Section 4 on page 71 below.

The second concern, and perhaps the harder to counter, is the inherent uncertainty which a characterisation analysis imports into the determination of whether a claimed invention is patentable or not. The characterisation process is inherently subjective in nature. This can be either a strength or a weakness, depending on your viewpoint.

Critics of the current approach in the US, which looks to distinguish a "technological" idea from an "abstract" idea,[85] claim that courts and patent examiners start from a presumptive categorical exclusion for claims directed to software and business methods, despite authority to the contrary.[86] On the other hand, since

---

84 In Australia, see *Microcell* (1959) 102 CLR 232 at 236 per Menzies J; *Research Affiliates LLC v Commissioner of Patents* (2014) 227 FCR 378 ("*Research Affiliates*") at 401 [107]; [2013] HCA 50; *D'Arcy v Myriad Genetics Inc* [2015] HCA 35; 258 CLR 334 ("*D'Arcy*") at [88], [94], [144].

85 Gene Quinn, "The Ramifications of Alice: A Conversation with Mark Lemley", *IP Watchdog*, 4 September 2014, <http://www.ipwatchdog.com/2014/09/04/the-ramifications-of-alice-a-conversation-with-mark-lemley/id=51023/> (21 June 2017) per Mark Lemley.

86 The US Supreme Court decision in *Diamond, Commissioner of Patents and Trademarks v Diehr et al.* (1981) 450 U.S. 175 ("*Diehr*") accepts the patentabilty of software, and the majority decision in *Bilski* made clear that business methods are patent eligible. Limiting

software interacts with the computer on which it is run and with various associated physical devices such as printers, scanners, and storage devices, one does not have to look far to find something physical enough to satisfy the inquirer that what is claimed is inherently patentable.

To properly understand how mischaracterisations might occur, it is apposite to recap the relevance of physicality to the inherent patentability inquiry.

D   *Physicality*

Physicality is a time-worn feature of patentable subject matter, whether it always existed or took an explicit form somewhere between the time of *Boulton v Bull*[87] and its expression in Morton J's "vendible product" test.[88] In Australian law the vendible product test continues to inform the patentable subject matter issue through its expanded re-characterisation in *NRDC*,[89] in particular the requirement of an artificially created state of affairs,[90] or something made by human action.[91]

In early Australian jurisprudence, the High Court held that "the ultimate end in view [of patentable subject matter] is the production or treatment of, or effect upon, some entity."[92] The modern reformulation of this "useful effect" test, although arguably derived from US law, first arose in *IBM v Commissioner of Patents*[93] where Burchett J stated that the difference between discovery and invention is distinguished by reference to "production of some useful effect".[94] As noted above, however, the focus on the end result is problematic.

---

categorical exclusions is the basis of statutory reforms proposed by various bar associations and groups such as AIPLA, IPO and NYIPLA: Dennis Crouch, "Sole Exceptions to Patent Matter Elibigibility," 29 June 2018, <https://patentlyo.com/patent/2018/06/exceptions-subject-eligibility.html> (2 July 2018).

87 All four judgments in *Boulton v Bull* include some form of physicality requirement. Eyre CJ held that a method to be patentable should be "embodied and connected with corporeal substances": at 667; Heath J limited patentability to two physical classes, namely "machinery [and] substances (such as medicines) formed by chemical and other processes": at 660; Buller J noted that a method was not patentable until its inventor had "carried it into effect and produced some new substance": at 663; and physicality is implicit in Rooke J's "new mode of construction": at 659, although it is acknowledged that his Honour did suggest that principles in the abstract might be patentable.

88 *Re GEC's Application* (1941) 60 RPC 1 at 4.

89 See for example *NRDC* at 277. The *NRDC* recharacterisation remains at the core of the patentabiity inquiry in Australia. See *D'Arcy* at [6].

90 The link between the two is expressed *R v Wheeler* (1819) 106 ER 392 (KB) at 395 as a requirement of "[s]omething of a corporeal and substantial nature, something that can be *made by man from the matters subjected to his art and skill*" (emphasis added).

91 *D'Arcy v Myriad Genetics Inc* [2015] HCA 35; 325 ALR 100; 115 IPR 1 at [6].

92 *Maeder v Busch* (1938) 59 CLR 684 at 705.

93 *International Business Machines Corporation v Smith, Commissioner of Patents* (1991) 33 FCR 218 ("*IBM Australia*").

94 *IBM Australia* at 224.

Any reference to an end result should be interpreted narrowly, as it was in *RCA Photophone* to "what practically achieved in physical fact".[95] On this basis, the end result is not the execution of the software on the computer but software created through the application of the inventive faculty. Whether this has any tangible physical manifestation is an entirely different question.

Perhaps the low point of support for physicality in Australia came in *Catuity* where Heerey J expressed doubt that such a requirement existed,[96] but in any event was satisfied that it was present along similar lines to the EU "any hardware" approach. The Full Federal Court in *Grant* has rejected this position,[97] explaining that what was meant in *NRDC* by an artificially created state of affairs was not a "useful effect" in the broad sense but a "useful *product*"[98] or a "concrete, tangible, physical, or observable effect".[99]

Physicality has also long been part of the US notion of patentability, whether accepted as a requirement[100] or as a "useful and important clue".[101] Physicality is impliedly required by even the broadest statements of patentability in that jurisdiction. For example, the US Supreme Court stated in *Chakrabarty*[102] that "anything under the sun made by man" is patentable; this was subsequently relied on for its allegedly expansive interpretation of patentable subject matter. At the very least the judgment implies a requirement of corporeality, however, in that some *thing* must exist "under the sun". Taking this view is further justified by the fact that the Court was in that case discussing only two branches of patentable subject matter under the US *Patents Act*, namely machines and manufactures.[103]

At the US Federal Circuit level, the *Freeman-Walter-Abele* (FWA) test[104] required that the claimed process be "applied to physical elements or process steps".[105] Despite this test being abandoned in *Alappat*,[106] the physicality of a claimed invention continued to inform the new requirement of a "useful,

95 *RCA Photophone*.
96 *Catuity* at 137.
97 That such physicality was not necessary was an argument put forward by the appellant.
98 *Grant* at 73 (emphasis added).
99 *Grant* at 70.
100 See for example *Mackay Radio v RCA* 306 US 86 (1939) which required a "novel and useful *structure*" for patentability.
101 *Bilski* at 3258 per Breyer J.
102 *Diamond v Chakrabarty* 447 US 309 (1980).
103 See *Bilski* at 3249 per Stevens J. His Honour notes that the quoted statement could not be taken to extend to processes in any event, as a process could not "be comfortably described as something '*made* by man'".
104 The *Freeman-Walter-Abele* test is short-hand for the approach suggested by the Federal Circuit in a series of three cases: *In re Freeman* 573 F.2d 1243 (1978); *In re Walter* 618 F.2d 766 (1980) and *In re Abele* 684 F.2d 902 (1982) ("*Abele*").
105 *Abele* at 907.
106 *In re Alappat* 33 F.3d 1526 (1994) ("*Alappat*").

concrete and tangible result".[107] A less stringent approach later took hold, according to which the claims need only covering something which *represented* measurable physical phenomenon.[108]

In the wake of *Bilski* this approach has fallen away. Following *Bilski*, the Supreme Court in *Mayo*,[109] *Myriad*,[110] and *Alice Corp.* have seemingly returned to first principles, endorsing a two-stage approach. The first stage of that approach involves asking whether the claims in question are directed to a judicially recognised exception to patentability, namely abstract ideas, laws of nature, or natural phenomena.[111] If so, the second stage requires consideration of whether the claims do significantly more than simply describe these things, that is, whether the additional elements transform the nature of the claim into a patent-eligible application of the judicial exception.[112]

In *Alice Corp.*, the two-step framework was applied to claims directed to a computer-implemented method, computer system, and computer-readable medium for mitigating settlement risk. Step one resulted in a finding that the claims were directed towards an abstract idea, namely the idea of an intermediated settlement.[113] The court then analysed the claims in search of "an inventive concept" to indicate "the patent in practice amounts to significantly more than a patent on the ineligible concept itself".[114] In the large majority of cases involving computer-related inventions since *Alice Corp.*, the US Federal Circuit has applied the two-step test to find the claims non-patentable.[115]

Whilst the status of the machine-or-transformation test from *Bilski* in the US is not entirely resolved following *Alice Corp.*, the Interim Guidance issued by the USPTO after *Bilski* sets out a number of "useful examples [which] are not intended to be exclusive or limiting. It is recognized that new factors may be developed, particularly for emerging technologies".[116] They are:

---

107 *Alappat* at 1544.
108 See also *State Street Bank & Trust Co v Signature Financial Group Inc* (1998) 149 F.3d 1368 ("*State Street*") where the "transformation of data, representing discrete dollar amounts" could amount to a useful effect. Cf. Robert Plotkin, "Computer Programming and the Automation of Invention," (2003) 7(2) *UCLA Journal of Law and Technology* <http://www.lawtechjournal.com/articles/2003/07_040127_plotkin.php> (19 April 2007): "The fact that a particular logical entity may represent a physical quantity or physical entity does not mean that the logical entity is itself physical."
109 *Mayo Collaborative Services v Prometheus Labs, Inc* 132 S. Ct. 1289 (2012).
110 *Association for Molecular Pathology v Myriad Genetics, Inc* 133 S. Ct. 2107 (2013) ("*Myriad*").
111 *Mayo Collaborative Services v Prometheus Labs, Inc* 132 S. Ct. 1289 (2012) at 1296–7 ("*Mayo*"); Affirmed in *Alice Corp* at 2355.
112 *Mayo* at 1297–8.
113 *Alice Corp* at 2355–57.
114 *Alice Corp* at 2355.
115 United States Patent and Trademark Office, "Patent Eligible Subject Matter: Report on Views and Recommendations from the Public" (July 2017) at 12.
116 United States Patent and Trademark Office, "Interim Guidance for Determining Subject Matter Eligibility for Process Claims in View of Bilski v Kappos" (2010) 75(143) *Federal Register* 43922) ("USPTO Interim Guidance 2010") at 43924.

A. Whether the method involves or is executed by a particular machine or apparatus. If so, the claims are less likely to be drawn to an abstract idea; if not, they are more likely to be so drawn. ...

B. Whether performance of the claimed method results in or otherwise involves a transformation of a particular article. If such a transformation exists, the claims are less likely to be drawn to an abstract idea; if not, they are more likely to be so drawn. ...

C. Whether performance of the claimed method involves an application of a law of nature, even in the absence of a particular machine, apparatus, or transformation. If such an application exists, the claims are less likely to be drawn to an abstract idea; if not, they are more likely to be so drawn. ...

D. Whether a general concept (which could also be recognized in such terms as a principle, theory, plan or scheme) is involved in executing the steps of the method. The presence of such a general concept can be a clue that the claim is drawn to an abstract idea.[117]

It seems that although not the sole test, the machine-or-transformation test will continue to play a role in the test for patentable subject matter. The notion of an application of a law of nature, or abstract idea, referred to in the second part of the two-step test, also involves the concept of physicality. So physicality continues to be an important concept in US patentable subject matter doctrine.

It is in EPO jurisprudence where the physicality of claims is arguably given the most lax interpretation. Early German interpretations of technical character led to a requirement that the claimed invention demonstrate some causal involvement with physical forces.[118] A different tack has been taken at the EPO, however. In *Vicom*[119] the EPO held that an alleged invention would be impermissibly "abstract so long as it is not specified what physical entity is *represented*".[120] This suggests an *Alappat*-style approach to physicality.[121] However, more recent EPO case law goes further than that, advocating an "any hardware" approach according to which the recitation of a physical feature in the claims will be sufficient to clear the patentability hurdle.[122] This is the point at which the

---

117 USPTO Interim Guidance 2010 at 43925–43926.

118 German patent law requires a "plan-conformant activity of using controllable natural forces to achieve a causally overseeable success which is, without mediation by human reason, the immediate result of controllable natural forces": Bundesgerichthof (German Federal Court of Justice) 22 June 1976, XZB 23/74 "Dispositionsprogramm". A partial English translation is available at <http://swpat.ffii.org/vreji/papri/bgh-dispo76/> (13 April 2007).

119 *Vicom/Computer-Related Invention* T208/84 [1987] EPOR 74 ("*Vicom*").

120 *Vicom* at 79.

121 Cf. *Fujitsu (No 1)* at 519 per Hearing Officer Haselden, who found that the representation of a technical artefact was insufficient to amount to a technical advance.

122 For example see T258/03 *Hitachi/Auction Method* [2004] EPOR 55 ("*Hitachi*") at [4.5] where technical character could be implied by "the physical features of an entity", which might even be satisffed by "activities which are so familiar that their technical character tends to be overlooked, such as the act of writing using pen and paper": at [4.6].

approach adopted by the UK courts differs from the EPO, where the recitation of hardware in the claims may well be ignored as a "confusing irrelevance".[123]

The Australian approach in *Research Affiliates* and *RPL Central* has been to adopt a hybrid of US and UK developments. A subsequent patent office decision in *Aristocrat Technologies*[124] conveniently summarises the current approach as follows:

> I conclude that it is relevant to consider a range of matters. Without seeking to be exhaustive, these include:
>
> - there must be more than an abstract idea, mere scheme or mere intellectual information;
> - is the contribution of the claimed invention technical in nature;
> - does the invention solve a technical problem within the computer or outside the computer;
> - does the invention result in improvement in the functioning of the computer, irrespective of the data being processed;
> - does the application of the method produce a practical and useful result;
> - can it be broadly described as an improvement in computer technology;
> - does the method merely require generic computer implementation;
> - is the computer merely an intermediary or tool for performing the method while adding nothing of substance to the idea;
> - is there ingenuity in the way in which the computer is utilised;
> - does the invention involve steps that are foreign to the normal use of computers; and
> - does the invention lie in the generation, presentation or arrangement of intellectual information.[125]

It is clear that this approach does attempt to draw a distinction between abstract or intellectual concepts on the one hand and physical, technical solutions on the other.

E Mischaracterisation

Even assuming that physicality is an express requirement, this is not dispositive of whether a software-based invention will be considered patentable. The EPO approach makes clear that such a requirement can rapidly devolve into a mere form requirement. The physicality of software has been asserted in a number of ways, which are considered below.

Software creates a new machine: In the first significant statement on the topic of software patents anywhere in the world, the United States President's

---

123 *Gale* at 326.
124 *Aristocrat Technologies Australia Pty Limited* [2016] APO 49 (22 July 2016) ("*Aristocrat*").
125 *Aristocrat* at [35].

Commission of 1966 report noted, after a substantial period of consultation and consideration, that

> [i]ndirect attempts to obtain patents and avoid the rejection, by drafting claims as a process, or a machine or components thereof programmed in a given manner, rather than as a program itself, have confused the issue further and should not be permitted.[126]

Despite this, courts and tribunals in the US, UK, and Australia, when reviewing claims to computer programs, have found sufficient physicality for such methods in the execution of that software on computer. In the US, only three years after the President's Commission, the CCPA in *Prater* held the mental steps doctrine was inapplicable to software claims because the claims included a "disclosure of apparatus for performing the process wholly without human intervention".[127] This was further explained in *Bernhart* as follows:

> [I]f a machine is programmed in a certain new and unobvious way, it is physically different from the machine without that program; its memory elements are differently arranged. The fact that these physical changes are invisible to the eye should not tempt us to conclude that the machine has not been changed. If a new machine has not been invented, certainly a "new and useful improvement" of the unprogrammed machine has been.[128]

Similarly in *Musgrave*, the CCPA relied on the performance of the claimed steps "by the disclosed apparatus"[129] in holding that the mental steps exclusion did not apply. This interpretation was returned to by the Federal Circuit in *Alappat* wherein the Court held that

> programming creates a new machine, because a general purpose computer becomes a special purpose computer once it is programmed to perform particular functions pursuant to instructions from program software.[130]

This interpretation was relied on in *State Street* to hold that "the transformation of data … by a machine … constitutes a practical application of a mathematical algorithm"[131] and was thus a patentable invention.

Pre-EPC decisions in the UK relied on this characterisation to find software patentable. In *Slee & Harris* the Patent Appeals Tribunal held that a computer

---

126 Presidential Commission on the Patent System, *To Promote the Progress of Useful Arts in an Age of Exploding Technology* (1966).
127 *Prater (No 1)* at 1403.
128 *Bernhart* at 1400.
129 *Musgrave* at 893.
130 *Alappat* at 1545.
131 *State Street* at 1373.

when programmed amounted to "a machine which is temporarily modified".[132] In *Badger's Application* a contrast was drawn between the conceptual and non-patentable preparation of data on the one hand, and on the other the patentable "conditioning of the computing apparatus so as to control its operation, in consonance with corrections laid down by the data record requirements".[133] The same Tribunal in *Gever's Application* characterised the claims therein as "data processing apparatus ... constrained to work in a certain way by punched cards which are inserted into it".[134] On a slightly different tack, it was held in *IBM (UK)* that although claims to no more than a standard computer could not be patentable subject matter, "the claim on its true construction ... only covers a computer when it is programmed to produce the required result and does not cover a standard computer as such."[135]

This line of cases was relied on in the Australian case of *IBM (Australia)* to support the proposition that the claims to a method of generating a curve on a computer "involve[d] steps which are foreign to the normal use of computers".[136] It was further stated, as had been suggested in *Burroughs*, that such a position was consistent with *NRDC*.[137]

European jurisprudence initially rejected the "new machine" approach. For example, in the *IBM Twins* case[138] the EPO seemed to acknowledge that this alone is insufficient, holding it necessary that the computer program brings about "a technical effect which goes beyond the 'normal' physical interactions between the program (software) and the computer (hardware) on which it is run".[139] Requiring something beyond these 'normal' interactions is consistent with the underlying question which the technical contribution requires to be answered, namely, what has the applicant invented?[140] Whether this includes some aspect of hardware is dependent entirely on the particular claims in question.

More recent EPO case law has implicitly embraced this approach, however, with the Board in *Pension Benefits* noting "a computer system suitably programmed for use in a particular field ... has the character of concrete apparatus

---

132 *Slee & Harris* at 55.
133 *Badger's Application* at 476.
134 *Gever's Application* [1969] FSR 480 at 486. See also *Burroughs* at 449 "The programme in fact constrains the apparatus to function in a particular way as long as the apparatus embodies that programme".
135 *International Business Machines Corporation's Application* [1980] FSR 564 (*"IBM (UK)"*) at 568.
136 *IBM (Australia)* at 226.
137 In *Burroughs* at 449 it was held that where the method results "in some improved or modified apparatus, or an old apparatus operating in a novel way, with consequent economic importance or advantages in the field of the useful as opposed to the fine arts".
138 T1173/97 *IBM/Computer program product* [1999] OJ EPO 609 (*"IBM Twin 1"*) and T935/97 *IBM's Application* [1999] RPC 861 (*"IBM Twin 2"*).
139 *IBM Twin 1* at [13].
140 See for example *Diehr* at 193–194; *In re Grams* 888 F.2d 839 (1989) (*"Grams"*) at 839; *Abele* at 907; *Sarkar* at 1333. In the UK, the approach is the same, see *Aerotel* at [40], [46].

in the sense of a physical entity, man-made for a utilitarian purpose and is thus an invention."[141] Similarly, the Board held in *Hitachi* that "the presence of technical character ... may be implied by the physical features of an entity."[142] Finally, in *Microsoft* the Board held that "[t]he claimed steps ... provide a general purpose computer with a further functionality. ... Moreover, the computer-executable instructions have the potential of achieving [a] further technical effect of enhancing the internal operation of the computer."[143]

The new machine view of software might be, as Samuelson suggests, one which "has some merit as a matter of computer science".[144] It is wrong to focus on it in the context of the subject matter issue, however. One reason for this is that "the source program remains at all times outside the machine and separate from it."[145] Thus a modified computer is the *end result* of writing code, (optionally) compiling it, loading it into a computer and running it. It may even be a *useful* result. But as was suggested in *RCA Photophone* "[t]hat involves an incorrect use of the word 'result', ... [which is] properly used in this connection as referring, *not to the end, but to the means.*"[146] In other words, what the applicant has invented may be causally related to the execution of software running on a general purpose of the computer, but this execution is not what the applicant has invented. The relevant means by which this result is achieved is a logical algorithm or procedure for processing data, which may or may not be meaningfully limited in any sense by the hardware of the computer.[147]

Other similarly inappropriate approaches include reliance on the printout of a receipt as part of the business method in *Catuity*, or the display of a curve on the screen as part of a claimed curve algorithm in *IBM Australia*. To use the words of the US Supreme Court, these amount to nothing more than insignificant post-solution activities.[148]

---

141 T931/95 *PBS Partnership/Controlling pension benefits systems* [2002] EPOR 522 ("*Pension Benefits*") at 530.

142 *Hitachi* at [4.5].

143 T424/03 *Microsoft/Clipboard formats I* [2006] EPOR 39 ("*Microsoft*") at [5.2].

144 Samuelson, "Benson Revisited", above n 59 at 1045, n 63.

145 Michael C Gemignani, "Legal Protection for Computer Software: The View From '79" (1979–1980) 7 *Rutgers Journal of Computers Technology and Law* 269 at 279.

146 *RCA Photophone* at 191, emphasis added.

147 The requirement of a meaningful limitation is an aspect of the machine-or-transformation test. See *In re Bilski* (2008) 545 F.3d 943 at 962: "the use of a specific machine or transformation of an article must impose meaningful limits on the claim's scope to impart patent-eligibility." (citing in support *Gottschalk v Benson* at 71–72).

148 See *Flook* at 590: "The notion that post-solution activity, no matter how conventional or obvious in itself, can transform an unpatentable principle into a patentable process exalts form over substance. A competent draftsman could attach some form of post-solution activity to almost any mathematical formula; the Pythagorean theorem would not have been patentable, or partially patentable, because a patent application contained a final step indicating that the formula, when solved, could be usefully applied to existing surveying techniques. n11 The concept of patentable subject matter under § 101 is not "like a nose of wax which may be turned and twisted in any direction...." *White v Dunbar*, 119 U.S. 47, 51."

Some commentators have sought to overcome such criticisms by defining software to cover only an "executable computer program."[149] This view might lend more support to the "new machine" argument, in that an executable computer program is causally close to the new machine. But it does not overcome the criticisms above, since the software is still, even on this definition, independent of the machine. It also glosses over the difficulties which arise in applying the distinction to interpreted languages where the source code is the executable code.[150] In any event, source code implementations would still face problems with the doctrine of equivalents and contributory infringement.[151]

Finally, a differently configured machine is not the inevitable (or at least only) result of writing a program. Gemignani describes other alternatives as follows:

> Just as a person who understands the [programming] language could use a program to reconstruct its algorithm, a person who has an adequate background in electrical engineering could use the program to build an electronic circuit which would carry out the program in conjunction with input and output devices. The program, to a suitable trained engineer, could be used as a circuit diagram or the blueprint for building special purpose hardware. Even if it is not part of a machine, the program could be used to construct a machine of specific design, that design being contained implicitly in the program itself. The program thus stands midway between the abstract solution to a problem and a machine which actually carries out that solution.[152]

Software as a structure on a carrier: Another way of reading in physicality into software claims has been by reference to the storage media on which the software is saved, compiled, transferred or loaded, be it hard drive, CD, USB stick, or other media. Until 2011, the accepted approach at the United States Patent and Trademark Office was:[153]

> When functional descriptive material is recorded on some computer-readable medium, *it becomes structurally and functionally interrelated to the*

---

149 Plotkin, above n 108. A similar approach is advocated in Mark A. Lemley & Eugene Volokh, "Freedom of Speech and Injunctions in Intellectual Property Cases" (1998) 48 *Duke Law Journal* 147 at 236.

150 See the discussion of interpreters and interpreted languages in Chapter 1 at 15.

151 Burk, above n 61 at 148–150.

152 Gemignani, above n 145 at 279.

153 United States Patent Office, "2106.01 Computer-related Nonstatutory Subject Matter" in Manual of Patent Examining Procedure (8th Ed, 2001), last revision July 2010 ("2010 MPEP 2106.01"). The approach changed after the Court of Appeals for the Federal Circuit decision in *CyberSource Corporation v Retail Decisions, Inc* (Fed. Cir. 2011, No 2009-1358) ("*CyberSource*"), which is properly seen as a refinement of the law post-*Bilski*.

medium and will be statutory in most cases since use of technology permits the function of the descriptive material to be realized.[154]

This indicates that so-called *Beauregard* claims,[155] being claims to "computer programs embodied in a tangible medium, such as floppy diskettes"[156] were considered patentable in the US, before *Bilski* at least. Also allowed were *Lowry* claims,[157] where what is claimed is a novel data structure on a carrier.[158] A less-tangible variation on these types of claims are directed to claims for the propagation of a signal through an intangible medium, however since the issuance of the USPTO interim guidelines in 2005,[159] and the *Nuijten* case,[160] it seems that such claims are no longer considered patentable on the basis that they do not fall within any of the four statutory classes of §101. Early indications after the adoption of the "machine or transformation test" by the Federal Circuit in *In re Bilski* suggested that *Beauregard* and *Lowry* claims remained patentable.[161] However, following *CyberSource*, the use of this type of claim is at an end. The court in *CyberSource*, considering the patent-eligibility of "a method for detecting credit card fraud",[162] looked beyond the form of the claims[163]

154 2010 MPEP 2106.01 (emphasis added). The 2010 MPEP drew a distinction between functional and non-functional descriptive material, which might broadly be considered to be a distinction between code and data.

155 *In re Beauregard* 53 F.3d 1583 (1995) (*"Beauregard"*). The Federal Circuit did not actually make a ruling in *Beauregard*, since the Commissioner for Patents conceded that such claims were statutory, and that the printed matter doctrine was not applicable. The patentability of such claims was thereafter generally accepted. See MPEP 2105.01 at section I.

156 *Beauregard* at 1584.

157 *In re Lowry* 32 F.3d 1579 (1994) (*"Lowry"*).

158 On first glance such claims are objectionable since the claims seem to be distinguished on the basis of their intellectual content, and run counter to the printed matter doctrine. However, as is the situation at the EPO, the printed matter doctrine bites back at the obviousness stage of the analysis, where "USPTO personnel need not give patentable weight to printed matter absent a new and unobvious functional relationship between the printed matter and the substrate" See 2010 MPEP 2106.01, citing in support *Lowry* at 1583–1584.

159 United States Patent and Trademark Office, "Interim Guidelines for Examination of Patent Applications for Patent Subject Matter Eligibility" *Official Gazette Notices*, 22 November 2005, <http://www.uspto.gov/web/offices/com/sol/og/2005/week47/patgupa.htm> (24 August 2010) ("USPTO Interim Guidelines 2005").

160 *In re Nuijten* 500 F.3d 1346 (2007) at 1352 wherein a 2–1 majority of the Court of Appeals for the Federal Circuit held that "transitory electrical and electromagnetic signals propagating through some medium, such as wires, air, or a vacuum … are not encompassed by any of the four enumerated statutory categories" and was therefore not patentable subject matter.

161 See *Ex parte Bo Li*, Appeal 2008-1213 (BPAI, 2008).

162 *CyberSource* at 17.

163 "Regardless of what statutory category ("process, machine, manufacture, or composition of matter," 35 U.S.C. § 101) a claim's language is crafted to literally invoke, we look to the underlying invention for patent-eligibility purposes." *CyberSource* at 17.

and found that what was claimed was in fact the same "abstract mental process"[164] the subject of an earlier claim.

According to the most recent USPTO Manual of Patent Examining Procedure (MPEP), the proper approach in this context is as follows:

> A claim whose [broadest reasonable interpretation ("BRI")] covers both statutory and non-statutory embodiments (under the broadest reasonable interpretation of the claim when read in light of the specification and in view of one skilled in the art) embraces subject matter that is not eligible for patent protection and therefore is directed to non-statutory subject matter. Such claims fail the first step (Step 1: NO) and should be rejected under 35 U.S.C. 101, for at least this reason. In such a case it is best practice for the examiner to point out the BRI and recommend an amendment, if possible, that would narrow the claim to those embodiments that fall within a statutory category.
>
> For example, the BRI of machine readable media can encompass non-statutory transitory forms of signal transmission, such as, a propagating electrical or electromagnetic signal per se. See *In re Nuijten*, 500 F.3d 1346, 84 USPQ2d 1495 (Fed. Cir. 2007). When the BRI encompasses transitory forms of signal transmission, a rejection under 35 U.S.C. 101 as failing to claim statutory subject matter would be appropriate. Thus, a claim to a computer readable medium that can be a compact disc or a carrier wave covers a non-statutory embodiment and therefore should be rejected under 35 U.S.C. 101 as being directed to non-statutory subject matter. See, e.g., *Mentor Graphics v. EVE-USA, Inc.*, 851 F.3d at 1294–95, 112 USPQ2d at 1134 (claims to a "machine-readable medium" were non-statutory, because their scope encompassed both statutory random-access memory and non-statutory carrier waves).[165]

The *Beauregard* and *Lowry* approaches are allowed at the EPO.[166]

Beauregard and *Lowry* claims are highly desirable from a litigation perspective since they do "not require any method steps to be performed to show infringement. Instead, direct infringement can be shown through the storage of the claimed data structure [or computer program] in a computer-readable medium".[167] On this approach, the "claim format is particularly helpful in

---

164 "[W]e find that claim 3 of the '154 patent fails to recite patent-eligible subject matter because it is drawn to an unpatentable mental process—a subcategory of unpatentable abstract ideas.": *CyberSource* at 9.

165 MPEP "2106 – Patent Subject Matter Eligibility" <https://www.uspto.gov/web/offices/pac/mpep/s2106.html> (17 July 2018).

166 See T258/03 *Hitachi/Auction Method* [2004] EPOR 55; T424/03 *Microsoft/Clipboard formats I* [2006] EPOR 39.

167 Daniel W McDonald, Robert A Kalinsky, and William D Schulz, "Software Patent Litigation" ABA Section of Litigation, Intellectual Property Litigation Committee Roundtable Discussion, April 2006 <http://euro.ecom.cmu.edu/program/law/08-732/Patents/SoftwarePatentLitigation.pdf> (20 August 2010) at 3.

proving that manufacturers and sellers of computer-readable media ... are direct infringers."[168]

The main problem with characterising software as a product on a carrier was aptly identified by the hearing officer in *Texas Instruments' Application*,[169] who noted that the carrier is a *"record* of the programme and not the programme itself". A similar approach was adopted in *Gale*. At first instance Aldous J held that "a ROM is *more than a carrier*, it is a manufactured article having circuit connections which enables to program to be operated."[170] On appeal their Lordships held that the ROM was "merely the vehicle used for carrying [a series of instructions]".[171] As such it was "it is convenient and right to strip away, as a confusing irrelevance, the fact that the claim is for 'hardware'."[172]

Hardware-software equivalence: From a purely functional perspective it is true that "a software process is often interchangeable with a hardware circuit."[173] It has also been argued that "[a] program's essential characteristic is the technical functionality represented in the instructions for the computer, and in this way is no different to specialised hardware."[174] As a result, it is tempting to conclude that "the software invention is no less tangible, useful, or worthy of patent protection."[175] Such a conclusion has often been espoused in case law.

In *Vicom* the EPO stated that "the choice between [hardware and software] is not of an essential nature but is based on technical and economic considerations which bear no relationship to the inventive concept as such."[176] The Board's decisions regularly emphasise this interchangeability.[177] Browne-Wilkinson VC in *Gale* made a similar assumption when he noted that "difficult cases can arise where *the computer program, whether in hardware or software*, produces a novel technical effect."[178]

168 McDonald et al., above n 168 at 3.
169 *Texas Instruments Inc's Application* (1968) 38 AOJP 2846.
170 *Gale* at 316–317.
171 *Gale* at 325 per Nicholls LJ.
172 *Gale* at 326.
173 David S. Bir, "The Patentability Of Computer Software After Alappat: Celebrated Transformation or Status Quo?" (1995) 41 *Wayne L Rev* 1531 at 1551, fn 121. See also Pamela Samuelson et al., "A Manifesto Concerning the Legal Protection of Computer Programs" (1994) 94 *Columbia Law Review* 2308 at 2319.
174 Advisory Council on Intellectual Property, *Report on a Review of the Patenting of Business Systems* (2003) <http://www.acip.gov.au/library/bsreport.pdf> (accessed 19 April 2007) at 8.4.1, summarising the arguments of IBM put forward at the European Patent Office International Academy Forum on the Protection of Computer-related and Business Model Inventions in November 1992.
175 Joseph R. Brown, "Software Patent Dynamics: Software As Patentable Subject Matter After State Street & Trust Co.," (2000) 25 *Oklahoma City University Law Review* 639 at 660.
176 *Vicom* at 80.
177 See *Vicom* at 79: "The technical means might include a computer comprising suitable hardware or an appropriately programmed general purpose computer"; *Vicom* at 80: "a program (be this implemented in hardware or in software)" and also referring to "a specified program (whether by means of hardware or software)".
178 *Gale* at 333 (emphasis added).

In the first instance decision of *Fujitsu (No 1)*, the Court stated a strong view that the two were entirely equivalent:

> In accordance with these principles, just as it would be possible to obtain a patent, considerations of novelty aside, for a faster chip or a more effective storage medium or a computer containing such a chip or storage medium, there is no reason in principle or logic why modification of the computer to achieve the same speed or storage increase by means of software should be excluded from protection. The fact that the advance is achieved in software rather than hardware should not affect patentability. To use in a slightly different context Nicholls L.J.'s words from *Gale's Application*, that would be to exalt form over substance. Similarly if a new process achieved by mechanical means would be patentable, there is no reason why the same process achieved by computer means should be any less patentable. If that is so, it does not matter whether the patent claims are drafted in terms of a process controlled by a computer, a computer when programmed in a particular way or a method of controlling a computer. In each case the substance of the invention is the same.[179]

In the US, a similar view was espoused in *Alappat*:

> In this field, *a software process is often interchangeable with a hardware circuit. Thus, the Board's insistence on reconstruing Alappat's machine claims as processes is misguided when the technology recognizes no difference and the Patent Act treats both as patentable subject matter.*[180]

As made clear above, however, physicality is has a long history in patent law. Despite regular emphasis upon the flexibility to adapt patent law to new conceptions of technology, this physicality requirement ought not be abandoned. The distinction between physical and intangible subject matter is considered in greater detail later in this book, but it is sufficient to say that in summary, the physicality of an invention will have impacts both on the commercialisation of the product, the structure of the market and most importantly, the nature of the creative process by which such products are created.

A similar criticism can be made of the claim that an extension of patentability to include software might be grounded in an *analogy* between the chemical process of *NRDC* and computer process claims.[181] There is a significant difference. Chemistry is a non-deterministic technology in which the inventive application of chemical principles to real-world problems must overcome an "infinity

---

179 *Fujitsu (No 1)* at 531 per Laddie J. This point was not considered by the Court of Appeal in when overturning his Honour's decision: *Fujitsu (No 2)*.
180 *Alappat* at 1583 per Rich J (emphasis added).
181 See *IBM (Australia)*.

of permutations" which can only be overcome by extensive field testing.[182] Software is not.

2.3   Complexity

Compared to the industrial-era machines which the patent system was originally designed to protect, a software product is a highly complex construction. Whilst a physical machine may be made of thousands of individual components, a large computer program (such as an operating system) may contain millions of components.[183] The fact that each individual component may potentially be patented makes infringement possible on a massive scale. Such a software product as just described could, by definition, potentially infringe millions of patents.

Another aspect of this high component-to-product ratio is that the patentability of such vast numbers of components can lead to an administrative nightmare for national patent offices. The USPTO originally put up stiff resistance to the introduction of software patents on the basis that allowing software patents would leave the patent office unable to handle the volume of applications which would follow. That is exactly what happened, with the USPTO in the period leading up to *Bilski* acknowledging that the backlog of patents had grown remarkably:

> [T]he volume and complexity of patent applications continues to outpace current capacity to examine them. The result is a pending – and growing – application backlog of historic proportions. Patent pendency … now averages more than two years. In more complex art areas, such as data-processing technologies, average pendency now stands at more than 3 years.[184]

The obvious result of such a backlog is the increased likelihood that individual patent examinations will not be carried out thoroughly, resulting in an across-the-board drop in patent quality. When doubtful patents are passed, the costs of pursuing litigation to see them overturned (usually at the risk of infringement proceedings) means that often such patents are successfully enforced against

---

182 Dukarich expressly notes chemical inventions as non-deterministic technologies. See Dukarich, above n 68, at 146–147.

183 Richard Stallman, "The Danger of Software Patents," Transcript of Speech Given at Cambridge University, March 2002, <http://www.cl.cam.ac.uk/~mgk25/stallman-patents.html> (13 September 2010) at [90].

184 United States Patent and Trademark Office, *Performance and Accountability Report for Fiscal Year 2005*, <http://www.uspto.gov/web/offices/com/annual/2005/2005annualreport.pdf> (12 July 2010) at 4. In 1989, the number of total applications pending was 222,755, but by 2005 this number had reached 885,002: United States Patent and Trademark Office, "Fiscal Year 2009: Patent Applications Pending Prior to Allowance" <http://www.uspto.gov/web/offices/com/annual/2009/oai_05_wlt_03.html> (12 July 2010). As at 30 June 2017, the backlog of unexamined applications stood at around 540,000 applications, and the average pendency for software-related applications stood at 28.5 months: see United States Patent Office, "Performance and Accountability Report – Fiscal Year 2017" <https://www.uspto.gov/sites/default/files/documents/USPTOFY17PAR.pdf> (17 July 2018).

competitors. Thus, the combined effect of the backlog and resultant grant of patents of dubious quality gave rise to regular criticisms in that period that the patent system in the US was "broken."[185]

2.4 Reuse

It will be recalled from Chapter 1 that the field of software development has advanced in an incremental fashion, from machine code to assemblers, assemblers to compilers and interpreters, and so on. Although "most inventions represent improvements on some existing article, process or machine,"[186] reuse lies at the very heart of software development. This creates unique challenges, which are outlined below.

A *Software innovations are largely cumulative*

Low barriers to entry in software development mean the market is massively decentralised. With the cheapest computer and an Internet connection, it is possible for anyone to get everything they need to create software. This decentralisation leads to independent repetition, "because programmers use similar, if not identical, software and hardware tools to tackle common needs."[187] Part of the reason for the similarity in tools is the influence of lower layers of the abstraction stack establishing a common framework for development at the upper levels.

This repetition, in combination with an emphasis on familiarity to users, and the ethos of sharing developed in the early days of computer science, "[p]rogrammers commonly adopt software design elements – ideas about how to do particular things in software – by looking around for examples or remembering what worked in other programs."[188] As such, most new developments of the technology tend to be cumulative, grain-sized innovations.

---

185 "Over the past several years, in newspaper articles and at hearings held by the Federal Trade Commission (FTC) and National Academy of Sciences (NAS), industry executives have complained in growing numbers that the patent system is broken." James Bessen and Michael J. Meurer, "Chapter 1: The Argument in Brief" *Patent Failure: how judges, bureaucrats, and lawyers put innovators at risk* (2008) at 3. The FTC and NAS hearings, and the reports produced as a result are discussed in Carl Shapiro, "Patent System Reform: Economic Analysis and Critique" (2004) 19 *Berkeley Technology Law Journal* 1017; Kurt M. Saunders, "Reforming the Patent System: Two Proposals" (2005) 4(1) *The Technology Report* <http://www4.ncsu.edu/~baumerdl/The%20Technology%20Report/Revised.Reforming.the.Patent.System.pdf> (13 July 2010). On the "broken" state of the patent system see also Adam B. Jaffe & Josh Lerner, *Innovation and its discontents: how our broken patent system is endangering innovation and progress and what to do about it* (2004).

186 *General Electric Company v Wabash Appliance Corporation* 304 U.S. 364, 368 (1938).

187 Ben Klemens, "New legal code: copyrights should replace software patents" (2005) *IEEE Spectrum Magazine*, <http://spectrum.ieee.org/computing/software/new-legal-code> (18 September 2010).

188 See Samuelson et al., above n 173 at 2330–2331.

The cumulative nature of software development has been consistently under-emphasised by courts and patent offices, meaning that large rewards (in the form of patent licences) are available as the result of minimal advances. The lure of large profits thus encourages opportunists to obtain and enforce broad patents at the expense of innovation in the field.[189]

B   *The software industry relies on openness*

From its origins in the computer science departments of universities, openness and collaboration has had an influential effect on the state of the art in software. At the earliest stages, computer scientists shared the source code of software innovations, either directly or through publication in freely available journals. This ethos of openness continued even as commercial players began to get involved. For example, the early history of the Unix operating system involved an open collaboration between AT&T and the University of California at Berkeley.[190]

Perhaps the most important form of openness in modern software is the Free and Open Source Software (FOSS) model, which provides a mechanism for global, collaborative innovation based on principles of freedom and sharing.[191] FOSS represents an important subset of the software industry because it embraces an alternative path to innovation and global collaboration which furthers the public interest in the development and disclosure of new technologies – the same goal claimed by the patent system.[192]

The continued importance of openness to innovation in software development is at odds with the award of monopoly rights. That software patents increase innovation in the field is also challenged by impressionistic evidence of the US software industry, which suggests that innovation levels were already high before software patenting began and have not increased since.[193]

FOSS developers face the same problems as all software developers, but there are ways in which FOSS projects are more vulnerable. Firstly, FOSS development is usually done by volunteers and by small service-oriented companies who make

---

189 See William W. Fisher III, "Intellectual Property and Innovation: Theoretical, Empirical, and Historical Perspectives" (2001) 37 *Industrial Property, Innovation, and the Knowledge-based Economy*, Beleidsstudies Technologie Economie, at 18.

190 For a more detailed history of the BSD project, see Marshall K. McKusick, "Twenty Years of Berkeley Unix: From AT&T-Owned to Freely Redistributable" in Chris Di Bona and Sam Ockman (eds) *Open Sources: Voices from the Open Source Revolution* (O'Reilly & Associates, 1999).

191 For more on the nature of FOSS, see Kenneth Wong and Phet Sayo, "Free/Open Source Software: A General Introduction," *UNDP-APDIP*, 2004 <http://hdl.handle.net/10625/50702> (12 July 2018).

192 "Accelerating the moment at which knowledge is widely available is consistent with patent policy's design to bring inventions into the public domain for the public benefit.": Robin Feldman, "The Open Source Biotechnology Movement: Is It Patent Misuse?" (2004) 6(1) *Minnesota Journal of Law, Science & Technology* 117 at 120.

193 Fisher above n 189 at 24–25.

their money out of customisation and administration of the software on customer projects.[194] Thus the problems of SMEs discussed above are particularly
relevant to the ongoing viability of FOSS projects. The problem is multiplied
since FOSS projects generally lack a central body to fund and harvest a large
patent portfolio, or to negotiate licensing agreements to cover project participants en masse.

Further, FOSS projects harness the collaborative power of the Internet, which
allows co-operation on projects on a global scale exceeding even that of the software industry monoliths.[195] The geographical distribution of FOSS project participants, together with the ability of any party to "fork" the code,[196] highlights
that no central body has ultimate control. Sometimes a project will have one or
more leaders,[197] or a non-profit organisation which holds the intellectual property and hosts development services,[198] but generally speaking, the levels of participation and responsibility for various aspects change from time to time. This
may be the key to the success of these projects, but the downside is that legal
responsibility for patent infringement is likely to be directed at individual con-
tributors or users of the software. It has been argued that the difficulty of pursuing a distributed target makes enforcement less likely,[199] but many FOSS
advocates feel that patents represent the biggest threat to this form of software
development.[200]

---

194 Empirical evidence gathered in relation to Embedded Linux supports this view, with 139
of 259 survey participants working for organisations with less than 50 people, and 81 participants working for companies with 200 or more staff. See Joachim Henkel and Mark
Tins, "Munich/MIT Survey: Development of Embedded Linux" (Institute for Innovation
Research, Technology Management and Entrepreneurship, University of Munich, 10 May
2004) <http://www.linuxfordevices.com/files/misc/MunichMIT-Survey_Embedded_
Linux.pdf> (5 September 2011) at 7.

195 Russ Roberts, "Why Linux is Wealthier than Microsoft" *BusinessWeek Online*, 19 November 2003 <http://yahoo.businessweek.com/technology/content/nov2003/
tc20031119_9737.htm> (18 September 2010).

196 Code forking is a consequence of the liberal modification and redistribution rights given in
OSS licences. If any one party is unhappy with the direction a project is taking, they can
take a copy of the source code and use it to launch their own project. See Wikipedia,
"Code Forking" <http://en.wikipedia.org/wiki/Code_forking> (18 September 2010).

197 For example, Linus Torvalds is the project leader for the Linux kernel project, and Guido
van Rossum is the leader of the Python language project.

198 For example the Apache Software Foundation (see <http://apache.org>) or the Mozilla
Foundation (see <http://mozilla.org/>) host the project website, version control and
mailing lists which allow project collaboration in addition to holding IP rights in the
project code.

199 Peter Williams, "Patent threat to open source is limited" *vnunet.com*, 8 September 2004
<http://www.pcw.co.uk/vnunet/news/2125807/patent-threat-open-source-limited>
(18 September 2010). Cf. Robert Jaques, "RIAA Launches P2P file sharing legal blitz"
vnunet.com, 19 November 2004 <http://www.v3.co.uk/vnunet/news/2126208/riaa-
launches-p2p-file-sharing-legal-blitz> (18 September 2010).

200 See for example Charles Babcock and Larry Greenemeier, "Open Source Stress" *Information Week*, 9 August 2004 <http://www.informationweek.com/story/showArticle.jhtml?
articleID=26806464> (17 July 2018).

The threat of FOSS projects being shut down by the spectre of patent litigation is a real one. In a study undertaken by Open Source Risk Management, it was found that the Linux kernel potentially infringed 283 US software patents which had yet to be tested by the courts.[201] Stallman has estimated that the Linux kernel represents only 0.25% of a complete GNU/Linux operating system,[202] suggesting between 30,000 and 300,000 possible infringements overall.[203] Although a "a third of the patents are owned by Linux backers, including Hewlett-Packard, IBM, Novell, and Oracle, which are unlikely to assert claims," this is nonetheless cause for concern.[204] The controversial decisions by some GNU/Linux distributors, including Novell, Xandros, and Linspire, to enter into a "Patent Cooperation Agreement" with Microsoft illustrates that many feel the threat is very real.[205]

2.5   Summary

It has been shown that abstraction, or abstractness, and context, or characterisation, form the primary dimensions of the software patent problem. The abstract nature of this subject matter holds the key to understanding why software should not be patentable. But attempts to address abstractness, through appeals to traditional exclusions such as mental steps, intellectual information, and abstract ideas, have faltered in the absence of a strong theoretical foundation. Case law demonstrates that software has been regularly mischaracterised, and its context ignored, so as to avoid difficult ontological questions about the nature of a particular set of software-related claims.

It has also been shown how complexity and reuse give rise to their own practical problems. The complexity of software gives rise to an over-supply of

---

201 Dan Ravicher, "Position Paper: Mitigating Linux Patent Risk" *Open Source Risk Management*, 2 August 2004 <http://www.osriskmanagement.com/pdf_articles/linuxpatentpaper.pdf> (4 September 2011) at 1.

202 See for example, Richard Stallman, "Patent absurdity" *The Guardian*, 20 June 2005 <http://www.guardian.co.uk/technology/2005/jun/20/comment.comment> (4 September 2005).

203 Richard Stallman, above n 202.

204 For example, the preamble to the GNU General Public Licence has for a decade recognised that "any free program is threatened constantly by software patents". See Free Software Foundation, "GNU General Public License" 16 February 1998 <http://www.gnu.org/copyleft/gpl.html> (18 September 2010).

205 See for example Microsoft Legal and Corporate Affairs, "Patent Cooperation Agreement – Microsoft and Novell Interoperability Collaboration" *Microsoft.com*, 2 November 2006 <http://www.microsoft.com/about/legal/en/us/IntellectualProperty/IPLicensing/customercovenant/msnovellcollab/patent_agreement.aspx> (4 September 2011); Rich Lehrbaum, "Linspire, Microsoft in Linux-related deal" *DesktopLinux.com*, 13 June 2007 <http://desktoplinux.com/news/NS9642338710.html> (4 September 2011); Xandros Inc, "Microsoft, Xandros Broad Collaboration Agreement Extends Bridge Between Commercial Open Source and Microsoft Software" on *xandros.com*, 4 June 2007 <http://www.xandros.com/news/press_releases/xandros_microsoft_collaborate.html> (4 September 2011).

patentable components which threaten both to overload patent offices and increase the probability that an independently developed product infringes multiple patents. The centrality of reuse to the software development process operates to reduce the grain size of software innovations and increase the likelihood of infringement. Reliance on reuse also means that the software industry's culture of openness, and alternative innovation paradigms such as FOSS, stands to be greatly disrupted by the award of monopoly rights.

Any proposed solution of the software patent problem will therefore need to address these issues.

3   How does the award of patents affect the software industry?

In addition to the problems which the nature of software causes patent law, there are a number of other considerations which bear on whether software ought to be patentable. These are discussed below.

3.1   Poor documentation means more trivial patents

Programmers are notoriously lousy at documenting the programs they write. This is perhaps because source code *is* documentation, or perhaps because compared to the active problem solving involved in writing code, writing documentation is just plain boring.[206] Even when documentation is properly written, however, the ubiquity of computers, the global nature of the industry, and the absence of any centralised knowledge repository (such as industry journals) means that it is very difficult to adequately capture the state of the art in documentary form at any point in time.[207] The absence of documentation presents serious problems to the patent examination process, as without a complete prior art base it is impossible for patent examiners to adjudge whether an alleged invention is truly novel and whether such an advance would have been obvious to a professional working in the field (the non-obvious requirement).

---

206 For a representative range of views on the issue, see "Why Do Programmers Hate Documenting?" <http://discuss.fogcreek.com/joelonsoftware1/default.asp?cmd=show&ixPost=35336&ixReplies=61>

207 As noted in the Introduction, in 2008 there were around 1.3 million computer systems software engineers and programmers in the US. That number was projected to increase by 21%, to around 1.6 million, by 2018: United States Bureau of Labor Statistics, "Computer Software Engineers and Computer Programmers" in *Occupational Outlook Handbook, 2010–11 Edition* <http://www.bls.gov/oco/ocos303.htm> (3 September 2011). Based on 2002 or later data on the relative size of the software industries in Brazil, China, India, Ireland, Israel, Japan, and Germany, it is estimated that the number of programmers in those countries would be in the region of 660,000. See Asish Arora and Alfonso Gambardella, "The Globalization of the Software Industry: Perspectives and Opportunities For Developed and Developing Countries" (National Bureau of Economic Research Working Paper 10538, June 2004) <http://www.nber.org/papers/w10538> (18 September 2010).

The risk is that trivial advances over the state of the art will regularly be elevated to the status of patentable invention.[208]

3.2   The software market moves too fast for the patent regime

A patent grant involves the award of monopoly rights to individual entities, typically for a 20-year period. This is an eternity in the fast-paced software industry, where a typical product lifecycle is around three to five years.[209] Because patents are construed as broadly as the language of their claims will allow, however, patent holders may assume control over technological developments far beyond that contemplated at the time of grant.

Even the application process is too slow. As noted above, the waiting period for the issue of a software patent is over two years.[210] The application will remain secret for some, if not all, of that period.[211] Given the independent repetition which characterises the industry, it becomes likely that entirely independent yet infringing software could be designed, implemented, brought to market and even replaced by a second generation before any knowledge of a relevant patent becomes available. In this scenario, the product's developer reaps none of the benefit of the patent and bears the entire burden of establishing a successful market for their product. But the patent holder is able to hold that developer's business hostage until royalties and/or a suitable licence are negotiated. Such royalties are an unwarranted tax on independent innovation.

---

208 This is the criticism typically leveled at the Amazon 1-click patent, which many developers consider to be an obvious application of cookie technology. For an enlightening discussion of the controversy, see Tim O'Reilly, "Ask Tim: Open Source, Patents & O'Reilly" *O'Reilly Network*, 28 February 2000 <http://oreilly.com/pub/a/oreilly/ask_tim/2000/amazon_patent.html> (18 September 2010). The Amazon 1-click patent is due to expire in 2017. See for example Ian Morris, "Amazon's Multi-Billion Dollar Patent Expires in 2017" <https://www.forbes.com/sites/ianmorris/2017/01/02/amazons-multi-billion-dollar-patent-expires-in-2017/> (21 August 2017).

209 For an overview see Wikipedia, "Moore's Law" <http://en.wikipedia.org/wiki/Moore%27s_law> (18 September 2010). For a more in-depth analysis, see Ilkka Tuomi, "The Lives and Death of Moore's Law" (2002) 7(11) *First Monday* <http://firstmonday.org/htbin/cgiwrap/bin/ojs/index.php/fm/article/view/1000/921> (18 September 2010).

210 See the discussion of the backlog at the USPTO on pages 62–63 above. Note that for Tech Center 2100 (Computer Architecture, Software & Information Security), the pendency for *first action* on a patent application at the USPTO for the most recent fiscal year is is 21.3 months, with total average pendency standing at 28.5 months. See United States Patent Office, "Performance and Accountability Report – Fiscal Year 2017" <https://www.uspto.gov/sites/default/files/documents/USPTOFY17PAR.pdf> (17 July 2018).

211 An ordinary application to a national patent office will not be published until the patent is granted. However, if it is an international application under the terms of the *Patent Cooperation Treaty*, [1980] ATS 6 (entered into force on 24 January 1978), Article 21(b) requires that, subject to certain exceptions, the international publication is to be "effected promptly" after 18 months.

3.3 Network effects amplify the power of patent holders

Network effects are common in software circles, where having a trained base of users for a particular product creates barriers to entry for competitor products. A network effect refers to "the social advantages that arise when all of the users of a particular type of technology adhere to the same standards and thus can share their work and move easily between machines and businesses".[212] A simple example is the Microsoft Word document, which is a de facto standard for storing and exchanging electronic documents.

Network effects may also arise as a result of the adoption by the industry of a particular standard. For example, the TCP/IP protocol is the standard for the transmission of digital information over computer networks, particularly the Internet. Standards are generally used to encourage interoperability and collaboration in a technologically optimal way.

Network effects amplify the ability of their owners to control a market by removing the bargaining power of competitors. For example, incorporating a patented technology into a standard allows the patent holder to enforce a licence fee from not only all competitors needing their products to interoperate with the standard, but also users of competing products.[213] This is because the patent cannot be "invented around" without deviating from the standard.[214]

3.4 Programmers hate software patents

If patent law's social contract is working as it should, then one would expect that those most likely to benefit from the disclosure of new technologies, namely programmers, would value patents as an important source of information. In fact most programmers believe that software patents "significantly hamper their work".[215] The League for Programming Freedom, for example, has been arguing for their abolition since at least 1991.[216] The voices of key of FOSS

---

212 Fisher above n 189 at 25.
213 Technically, use of the product would fall within the scope of the patent holder's right to exploit the invention. It is assumed however that a patent holder is unlikely to sue its own customers.
214 It is to get around network effects that patent pools are often created. Depending on the market power of those who participate in the pool, such pools may have either pro-competitive or anti-competitive effects. See Steven C. Carlson, "Patent Pools and the Antitrust Dilemma" 16 *Yale Journal on Regulation* 359 (1999); Jeanne Clark et al. "Patent Pools: A Solution to the Problem of Access in Biotechnology Patents" (2001) 20(4) *Biotechnology Law Report* 607; Robin Feldman, "The Open Source Biotechnology Movement: Is It Patent Misuse?" (2004) 6(1) *Minnesota Journal of Law, Science & Technology* 117 at 166–167.
215 Fisher above n 189 at 25.
216 See League for Programming Freedom, "Against Software Patents" 28 February 1991 <https://web.archive.org/web/20150329103351/http://www.progfree.org/Patents/against-software-patents.html> (17 July 2018).

figures are unanimous in arguing for the abolition of software patents.[217] There also is empirical[218] and anecdotal evidence[219] to suggest that proprietary software developers would happily do without software patents. Why is this?

Despite the prevalence in the public consciousness of large US firms such as IBM, Microsoft and Oracle, the global software industry is largely populated by small-to-medium enterprises (SMEs).[220] In Australia, for example, a 2005 study found that 99.6% of Australian computer software and services companies contained less than 100 employees.[221] These small players are particularly vulnerable to the problems software patenting creates. Their limited budgets mean that they cannot afford license fees, the costs involved in challenging validity or defending against infringement. It is they who feel the pain of software patenting most sorely.[222]

In contrast, software giants are generally able to limit the likelihood of infringement proceedings by maintaining large patent portfolios. These portfolios are reminiscent of the "mutually assured destruction"[223] policy of the superpowers in the Cold War, in that any would-be litigant is discouraged from initiating proceedings lest they should be hit with a counterclaim for

217 See Richard Stallman, "The Danger of Software Patents" 2004 Cyberspace Law and Policy Seminar (audio recording), Sydney, 14 October 2004 <http://www.bakercyberlawcentre. org/2004/talks/Stallman_1_small.ogg> (18 September 2010) (Part 1) and <http://www. bakercyberlawcentre.org/2004/talks/Stallman_2_small.ogg> (18 September 2010) (Part 2); Linus Torvalds and Alan Cox, "Open Letter on Software Patents from Linux Developers" 21 September, 2003 <http://www.effi.org/patentit/patents_torvalds_cox.html> (5 September 2011); Robert McMillan, "Torvalds joins in anti-patent attack" *Techworld*, 2 February 2005, <http://news.techworld.com/applications/3059/torvalds-joins-in-anti-patent-attack/> (18 September 2010); Bruce Perens, "Public Policy Area: Software Patents" <http://perens.com/policy/software-patents/> (18 September 2010); Bruce Perens, "The Monster Arrives: Software Patent Lawsuits Against Open Source Developers" <http://technocrat.net/d/2006/6/30/5032/> (18 September 2010).
218 See Effy Oz, "Acceptable Protection of Software Intellectual Property: A Survey of Software Developers and Lawyers." (1998) 34(3) *Information & Management* 161.
219 The selection of anti-software-patent quotes collated on this page read like a virtual who's-who of the US Software Industry. See Foundation for a Free Information Infrastructure, "Quotations on Software Patents," <http://eupat.ffii.org/vreji/quotes/index.en.html> (18 September 2010).
220 See Samuelson et al., above n 173 at 2377.
221 Centre for Innovative Industry Economic Research Consortium, "Executive Summary", *The Australian software industry and vertical applications markets: Globally competitive and domestically undervalued*, <http://www.dbcde.gov.au/__data/assets/pdf_file/ 0020/36461/Executive_summary.pdf> (5 September 2011).
222 On this point see Puay Tang, John Adams and Daniel Paré, "Patent protection of computer programmes" (Final Report submitted to the European Commission, 2001). See also Foundation for a Free Information Infrastructure, "Alliance of 2,000,000 SMEs against Software Patents and EU Directive," 16 September 2003, <http://eupat.ffii. org/papers/eubsa-swpat0202/ceapme0309/> (18 September 2010).
223 See Wikipedia, "Mutually Assured Destruction," <http://en.wikipedia.org/wiki/Mutually_ assured_destruction> (18 September 2010).

infringement themselves. However, software patenting nevertheless impacts negatively on them. Maintaining such a strategy consumes time and money:

> The time and money we spend on patent filings, prosecution, and maintenance, litigation and licensing could be better spent on product development and research leading to more innovation. But we are filing hundreds of patents each year for reasons unrelated to promoting or protecting innovation.[224]

Defensive portfolios also do little to discourage so-called patent trolls – companies who do not engage in software development and whose sole source of income depends on opportunistic enforcement of the patents they acquire.[225]

4   Is this a subject matter issue?

One point which emerges from the survey of EPC jurisprudence is that the EPO Board of Appeal considers the whole problem to be an inventive step issue, not a subject matter issue. For example, the Enlarged Board in G3/08 noted that:

> While the Enlarged Board is aware that this rejection for lack of an inventive step rather than exclusion under Article 52(2) EPC is in some way distasteful to many people, it is the approach which has been consistently developed since [the *IBM Twins*] and since no divergences from that development have been identified in the referral we consider it not to be the function of the Enlarged Board in this Opinion to overturn it.[226]

A similar attitude is evident in the US Federal Circuit decision in *Diehr*, who in rejecting the *Flook* "point of novelty" approach to determining patentable subject matter stated that

> [the question] of whether a particular invention is novel is wholly apart from whether the invention falls into a category of patentable subject matter.[227]

The majority in *Bilski* also seem to put much faith in the other requirements for patentability:

> The §101 eligibility inquiry is *only a threshold test*. Even if a claimed invention qualifies in one of the four categories, it must also satisfy "the

224 Robert Barr, Statement to the Joint Federal Trade Commission and Department of Justice Hearings on Economic Perspectives on Intellectual Property, Competition and Innovation, 28 February 2002, <http://www.ftc.gov/opp/intellect/barrrobert.doc> (5 September 2011).
225 For more on how patent trolls operate, see Wikipedia, "Patent Troll," <http://en.wikipedia.org/wiki/Patent_troll> (18 September 2010).
226 G3/08 *Programs for computers* (Opinion of 12 May 2010) at [10.13].
227 *Diehr* at 190. See also *Bernhart*.

conditions and requirements of this title," §101(a), including novelty, see §102, nonobviousness, see §103, and a full and particular description, see §112.[228]

Given the difficulty set out above, it may not be surprising that decision makes might wish to focus instead on a more technical question of whether a claimed invention is new. The answer to that was made clear by the UK Court of Appeal in *Aerotel*:

> Consider for instance the following:
>
> i   a claim to a book, e.g. to a book containing a new story the key elements of which are set out in the claim;
> ii a claim to a standard CD player or iPod loaded with a new piece of music.
>
> Everyone would agree that the claims must be bad – yet in each case as a whole they are novel, non-obvious and enabling. To deem the new music or story part of the prior art (the device of *Pension Benefits* and *Hitachi*) is simply not intellectually honest. And, so far as we see, the *Microsoft* approach, which discards that device, would actually lead to patentability.[229]

Christie and Syme identify a further reason why the inventiveness test provides an unsatisfactory work-around. The test for inventiveness takes place in the context of the relevant art.[230] When a computer program forms the substance of the claims, then the relevant art is likely to be computer programming. However, computer programming often involves the automation of processes in other areas. To use Christie and Syme's example,[231] when a mathematical algorithm is translated to computer-automated form, it will not be relevant that the

---

228 *Bilski* at 3221 per Kennedy J (emphasis added). Note that the USPTO's post-Bilski interim guidelines have picked up on this remark, and remind examiners "that § 101 is not the sole tool for determining patentability ... Therefore, examiners should avoid focusing on issues of patent-eligibility under § 101 to the detriment of considering an application for compliance with the requirements of §§ 102, 103, and 112, and should avoid treating an application solely on the basis of patent-eligibility under § 101 except in the most extreme cases": USPTO Interim Guidance 2010 at 43923–43924. Cf. *Bilski* per Stevens J – see n234 below.

229 *Aerotel* at [27].

230 Christie and Syme are discussing the Australian approach to inventive step, but their comments are equally applicable to the US and UK approaches. As to the UK and EPO, see *Pozzoli Spa v BDMO SA & Anor* [2007] EWCA Civ 588 (22 June 2007); T154/04 *Duns/Method of estimating product distribution* [2008] OJ EPO 46 at Reasons 5(G). Similarly the US approach is tested by reference to the "personal having ordinary skill in the art" §103 *Patent Act 1952* (US). For a more recent interpretation given to this standard see *KSR International Co. v Teleflex Inc., et al.* 550 U.S. 398 (2007).

231 See Andrew Christie & Serena Syme, "Patents for Algorithms in Australia," (1998) 20 *Sydney Law Review* 517 at 553.

algorithm itself has been in use in the field of mathematics for hundreds of years if it has not been used in computer programming before.

Further, any inventive step determination is hamstrung by the incomplete prior art due to the lack of documentation noted above. Whether the prior art can be sufficiently determined is not an issue which falls to be considered within the confines of the inventive step test, but would instead be an appropriate consideration in determining the inherent patentability of a category of inventions likely to suffer from this malady. This is because inherent patentability is the "first and most fundamental"[232] threshold which any patent application must surmount in the patentable subject matter inquiry. Whereas the other requirements focus on the technical aspects of the invention, the patentable subject matter inquiry is intended to establish the *prima facie* availability of patents for subject matters suitable for the award of patents.[233] As such, this inquiry can (and it is submitted should) address a broader range of considerations.

In other words, "the requirements of novelty, nonobviousness, and particular description [can not] pick up the slack."[234] That notions of newness and inventiveness may be touched on by the inherent patentability inquiry is not a new development – the history of excluding analogous uses from patentability at the very least confirms that.[235] The nature of inherent patentability as a threshold issue, subject as it is to the need to determine what the applicant has invented,[236] means that some consideration of the nature of the invention is required. To hold that the characterisation process unnecessarily overlaps with other requirements is to ignore the history and the purpose of the inquiry.

5   How should the analysis proceed?

From the above, a number of issues have been identified. First, the lack of a satisfying account as to why abstract ideas are non-patentable is of concern.

---

232 Advisory Council on Intellectual Property, *Patentable Subject Matter: Issues Paper* (2008) <http://www.acip.gov.au/library/Patentable%20Subject%20Matter%20Issues%20Paper. pdf> (accessed 13 November 2008) at 1.

233 Justine Pila, "Response to the Australian Government Advisory Council on Intellectual Property's Request for Written Comments on Patentable Subject Matter" *Oxford Legal Studies Research Paper No. 37/2008*, September 19, 2008 <http://papers.ssrn.com/ sol3/papers.cfm?abstract_id=1274102>.

234 *Bilski* at 3238, n 5 per Stevens J. His Honour also noted the "substantial academic debate, ...about whether the normal process of screening patents for novelty and obviousness can function effectively for business methods. The argument goes that because business methods are both vague and not confined to any one industry, there is not a well-confined body of prior art to consult, and therefore many "bad" patents are likely to issue, a problem that would need to be sorted out in later litigation.": at 3256 n55. This accords with the criticisms just levelled at software patents.

235 See *Losh v Hague* (1838) 1 Web Pat 202 (NP); *Harwood v Great Northern Railway* (1865) 11 ER 1488; *Muntz v Foster* (1843) 2 WPC 93. See also *Crane v Price* (1842) 134 ER 239.

236 See n 140 above.

Second, the lack of a theoretical underpinning for exclusions such as the mental steps doctrine has resulted in their erosion over time. This suggests the failure to properly articulate the reasons for excluding abstract ideas may lead to this exclusion being eroded. Given that this exclusion may be all "that is left to stand between all conceivable human activity and patent monopolies",[237] such a result would be disastrous.

Many of the problems reviewed in this chapter stem from the limitations of a narrow, economic/utilitarian approach to patent law theory, and thus to the patentable subject matter inquiry. Many of the issues explored do not fit neatly within such an analysis. Therefore if the software patent problem is to be resolved, a new approach is required.

Given the isomorphism between mathematics and software outlined in Chapter 1, one might expect the non-patentability of the former would be dispositive of the status of the latter. But the relationship between software and mathematics is poorly understood in patent law. The dominant approach in cases such as *Benson* has been through the lens of the 'algorithm', which assumes some software is mathematical, whereas some is not. Such a distinction is illusory.[238]

Although it is generally accepted that mathematics is not patentable, the reasons for this have never been identified. Mathematics is a discipline as old as humankind. Yet in the 400-year history of patent law it has received only tangential consideration. Given the closeness of the relationship between software and mathematics, the nature of mathematics and its patentability are questions of central importance to the patentability of software-related inventions. For that reason, it is proposed to pursue the patentability of software by reference to the patentability of mathematics.

6   Conclusion

This chapter has explored the parameters of the software patent problem, informed by the historical foundations of patentable subject matter jurisprudence and modern applications of that jurisprudence to this unique and confounding subject matter. The way in which courts and tribunals in every jurisdiction have been confounded in their attempts to compellingly explain why software should or should not be patentable leads to the conclusion that a new approach is required. It has been posited that the reason why software's patentability is a vexed question is because of its misunderstood relationship with mathematics, and that the nature of mathematics and the reasons for its non-patentability may lead to a solution.

---

237 *Bilski* at 3238, note 5 per Stevens J.
238 See Newell, above n 60 at 1024–1025 wherein Newell notes that "there is an underlying identity between the numerical and the nonnumerical realms that will confound any attempt to create a useful distinction between them".

The journey towards that solution must begin with a detailed consideration of the nature of mathematics. As will be seen in the next chapter, the avoidance of mathematics by patent lawyers may be an informed choice. There is much dispute, even among mathematicians, as to what mathematics is. It is not proposed to solve that question, but it is expected that by reviewing the disparity of theories as to the nature of mathematics, a theory of its non-patentability might be developed.

3   The nature of mathematics

Courts have used the terms 'mathematical algorithm,' 'mathematical formula,' and 'mathematical equation,' to describe types of nonstatutory mathematical subject matter without explaining whether the terms are interchangeable or different. Even assuming the words connote the same concept, there is considerable question as to exactly what the concept encompasses.[1]

Mathematics is an interesting topic to look at from a patent law perspective because it forms one of the longest standing and perhaps least understood exceptions to patentability. Although it is generally accepted that mathematics is to be excepted from patentability, the reason why is rarely considered. It seems to be merely a thus-far unchallenged assumption, or perhaps an intuition. One typical explanation of the mathematics exception suggests that mathematics "[tends] to fall in the category of laws of nature, discoveries, scheme and plans, and fine arts as opposed to useful arts."[2] In other words, a little bit of everything is at play. This chapter will look to both the philosophy of mathematics and the historical development of the patent system in order to find a sound theoretical basis for its non-patentability.

This chapter looks first at the historical and philosophical understandings of mathematics. It will be shown that whilst many theories as to the nature of mathematics have been advanced, no particular theory has come to dominate. It may be wondered what the value of such a seemingly dead-end inquiry might be. First, it is necessary foundational material by which to understand the nature of innovation in the field of the mathematical arts. Whilst no one theory dominates, all theories have value in that they illuminate an understanding of the activity of mathematics, and by extension, the nature of innovation in that field.

Second, it is only through an understanding of the various accounts of the nature of mathematics that explanations for its non-patentability can be analysed.

---

1 *AT&T Corp v Excel Communications Inc* 172 F.3d 1352 (Fed Cir, 1999) ("*AT&T*") at 1353.
2 Rocque Reynolds and Natalie P Stoianoff, *Intellectual Property: Text and Essential Cases* (2nd ed, Federation Press, 2005) at 268, n46.

It is to this end that the second half of the chapter is drawn. It will be shown how these lawyers' accounts of mathematics cannot embrace all theories of mathematics, and as a result, are unsatisfactory explanations.

1   A mathematician's account of mathematics

At the outset, it is important to have some understanding of what mathematics is. A starting point towards such an understanding might be to state that it is what mathematicians study, although such a definition does not advance things very far. Providing a definitive definition of mathematics is difficult, because the nature of mathematics has changed over time, as its use has become more widespread, and as developments in mathematics combined to shake the very foundations of mathematicians' conception of their field.[3]

Mathematicians themselves have defined their art in broad, poetic terms. For example, it has been said that mathematics is:

- "the substance of thought writ large";[4]
- "the queen of sciences";[5]
- "the art of reason";[6] and
- "a science of patterns".[7]

But such descriptions, although they provide some clues as to the nature of mathematics, lack the detail necessary to form a solid basis on which to build an argument. A more conventional legal approach is to start with a dictionary definition:

1   a group of related sciences, including algebra, geometry, and calculus, concerned with the study of number, quantity, shape and space, and their interrelationships by using a specialized notation
2   mathematical operations and processes involved in the solution of a problem or study of some scientific field.[8]

---

3   Simon McLeish, "Mathematics" in Kenneth McLeish (ed), *Bloomsbury Guide to Human Thought*, (Bloomsbury Publishing, 1993). See also "I.1 What is Mathematics About?" in Timothy Gowers et al. (eds), *The Princeton Companion to Mathematics* (Princeton University Press, 2008): "It is notoriously hard to give a satisfactory answer to the question, 'What is mathematics?' The approach of this book is not to try".
4   Leonard Peikoff, *Objectivism: The Philosophy of Ayn Rand* (Penguin Group, 1991) at 90.
5   Carl Friedrich Gauss, cited in Guy W Dunnington, Jeremy Gray and Fritz-Egbert Dohse, *Carl Friedrich Gauss: titan of science* (MAA, 2004) at 44.
6   William P. Berlinghoff, *Mathematics: The Art of Reason* (DC Heath & Company, 1968).
7   Michael D. Resnik, *Mathematics as a science of patterns* (Clarendon Press, 1997).
8   Elspeth Summers and Andrew Holmes (eds), *Collins Australian Dictionary and Thesaurus* (3rd ed, HarperCollins Publishers, 2004).

Now the many-faceted nature of mathematics begins to reveal itself. First there is an ontological definition: that mathematics is merely a grouping term for a number of related disciplines, with a common concern for number, quantity, shape and space, and their interrelationships. Such a definition raises a question. Is there a "unity of mathematics" which even these secondary concepts have in common? Hiding behind this definition are further questions about the nature of the objects within these fields of study – for example, what kind of objects are numbers? Are they mental constructions, or do they exist independently of us?

The second definition stands in contrast to the first in that it suggests that perhaps mathematics has no subject matter, it is merely something one *does*. By classifying mathematics as a type of science, this definition skirts around a number of epistemological questions about mathematics. In the natural sciences, knowledge is verified by empirical observation. But the content of mathematics is abstract "objects" like numbers, variables, and so on. Without validation through empirical observation, how can the truth of mathematical propositions be proved? What does it mean to say that something in mathematics is "known"?[9] It is clear that the answers to these epistemological questions may depend on the way in which the ontological questions are answered. This interdependence runs in both directions.[10] It is clear then, that such definitions do not provide an adequate account of the nature of mathematics by themselves.

More practically, mathematics can be described as "a process of thinking that involves building and applying abstract, logically connected networks of ideas. These ideas often arise from the need to solve problems in science, technology, and everyday life-problems."[11] At the heart of mathematics lies a process of abstraction – moving the focus away from the limitations of the immediate physical problem to be solved into an abstract model.[12] This process of abstraction is based on two aspects – simplification and generalisation. For example, a simplification may be used if we need to work out the area of one face of a plank of wood, as shown in Figure 3.1. To simplify its actual shape, one would smooth out all the bumps and variations along its edges and think of it as a rectangle, with four straight lines meeting at right angles. Although not exact, the level of accuracy attained will be more than sufficient for most purposes.

9 Some of these issues are touched upon in the account of the history of mathematics below.
10 Sam Butchart, *Evidence and Explanation in Mathematics* (PhD Thesis, Monash University, 2001) at 8.
11 American Association for the Advancement of Science, "The Nature of Mathematics" on *Benchmarks Online* <http://www.project2061.org/publications/bsl/online/ch9/ch9. htm#MathematicalWorld> (accessed 12 August 2018).
12 This should come as no surprise in consequence of abstraction having been identified as a key aspect of software, and the isomorphism between mathematics and software discussed in Chapter 1.

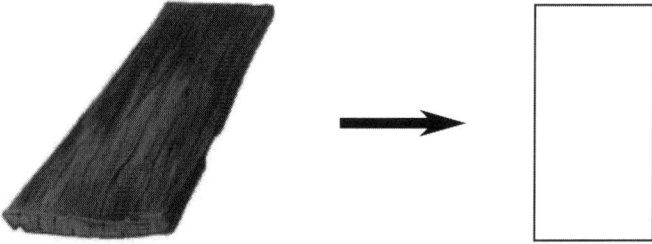

Figure 3.1 Abstraction of a real world object to a simple shape

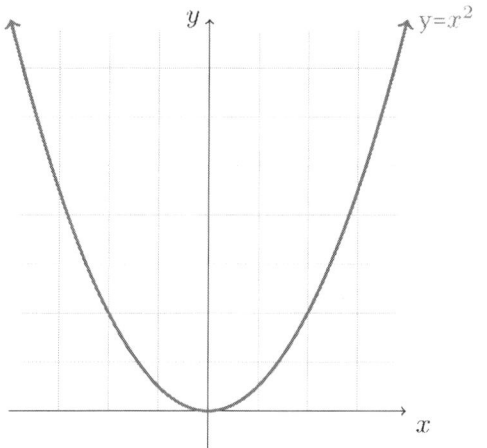

Figure 3.2 A parabola

A similar process might be followed for many planks of wood of similar dimensions. Here, the simplification might involve using average dimensions, informed by taking measurements of a small sample of the planks involved. There might be no exact correspondence between the measurements of actual pieces of wood and the idealised average piece. But this sort of simplification makes estimations and modelling more workable.

The process of abstraction allows mathematics to operate as "a kind of 'place' in which logical exactness can be 'applied' to the utmost degree".[13] Further, such abstract models often turn out to have applications beyond the original problem domain. For example, consider the parabola in Figure 3.2.

13 Evandro Agazzi, "The Rise of the Foundational Research in Mathematics," (1974) 27 *Synthese* 7 at 9.

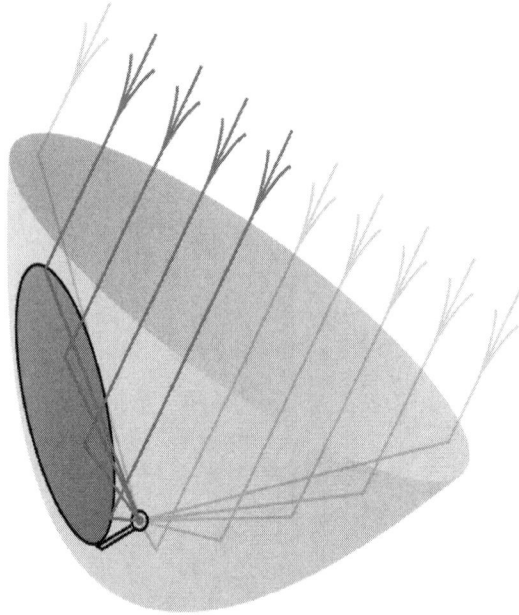

Figure 3.3 Geometry of an offset parabolic satellite dish

The reflective properties of the parabola have been known for a long time, allegedly as far back as the 3rd century BC when Archimedes used a series of reflecting mirrors arranged along a parabolic path to defend Syracuse from the Romans. The same reflective properties mean the shape is still used today in headlights and satellite dishes, the latter being shown in Figure 3.3.[14]

But the parabola has relevance beyond its reflective properties. In the 17th century, Galileo discovered through experimentation that the trajectory of an object travelling through the air had approximately the same shape. Isaac Newton later proved this to be the case. The parabola can also be used to model the shape of the main cable in suspension bridges, such as the Golden Gate bridge, the surface of a liquid confined to a container and rotated around the central axis, and the trajectory of a bouncing ball (the latter two being shown in Figure 3.4).[15]

14 Image by Wikimedia Commons user 'cmglee', <https://commons.wikimedia.org/wiki/ File:Off-axis_parabolic_reflector.svg> (29 June 2017). Licensed under the Creative Commons Attribution-Share Alike 3.0 Unported license.
15 Wikipedia, "Parabola" <http://en.wikipedia.org/wiki/Parabola> (30 October 2007). Picture of the rotating liquid by Matthew Trump, 26 February 2005, <https://en.wikipe dia.org/wiki/File:Parabola_shape_in_rotating_layers_of_fluid.jpg> (29 June 2017) licensed under the Creative Commons Attribution-Share Alike 3.0 Unported license. Picture of

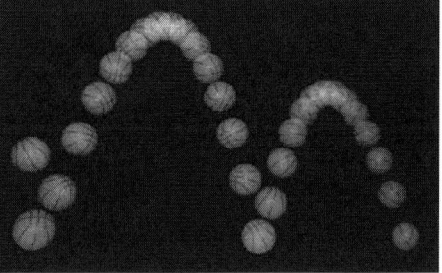

(a) Parabola in a liquid surface under rotation.   (b) Parabolic trajectory of bouncing ball.

Figure 3.4 Parabolas in nature

This demonstrates that often, mathematical abstractions not only owe their origins to the study of real-world problems, but also that many real-world problems have deep connections to each other, even in seemingly unrelated areas.

The recurrence of mathematical artefacts in various physical phenomena lends an air of authenticity to the objects on which mathematics is based. Yet the study of mathematics often deals in concepts which seem to have no direct relationship to the real world at all. Natural numbers, even very large numbers,[16] correspond to real things. But what about 2^{65536}? This number exceeds the largest estimation of physically observable phenomena in the universe.[17] Even geometry, a discipline long considered the only "true" mathematics due to its verifiability by human sensory experience, faces similar questions. A straight line may seem uncontroversial, but the straight line used in mathematics has no width, and as such cannot be drawn at all. The so-called straight line one might draw on a page or print on a printer is in reality a very thin rectangle.[18] Given it has no manifestation in physical form, does a straight line exist?

Such questions might have been restricted to Sunday afternoon debates between mathematicians at the local pub, were it not for their practical utility.

---

trajectory of bouncing ball by Michael Maggs, edited by Richard Bartz, <https://en.wikipe dia.org/wiki/Parabola#/media/File:Bouncing_ball_strobe_edit.jpg> (29 June 2017), licenced under Creative Commons Attribution-Share Alike 3.0 Unported Licence.

16 Consider 20,350,000,000,000. This number represents the upper bounds of how many bytes of email were sent in the year 2000. Taken from Peter Lyman and Hal R. Varian, "How Much Info? Internet" (School of Information Management and Systems, University of California, 2000) <http://www2.sims.berkeley.edu/research/projects/how-much-info/internet.html> (7 September 2011).

17 The estimate which the number exceeds is the total number of vibrations of all subatomic particles in the universe over its entire existence. See David Isles, "What Evidence is there that 2^65536 is a natural number?" (1992) 33(4) *Notre Dame Journal of Formal Logic* 465.

18 Even that is an approximation since if you look at it closely enough, it is unlikely to have straight edges.

For example, negative numbers have no direct correspondence to things in the real world. You cannot hold −1 orange in your hand. But the use of a negative numbers is commonplace in representing debts in accounting. Similarly, the so-called irrational numbers such as e, $\sqrt{2}$ and π have values which cannot be represented by a simple fraction,[19] yet they are essential for geometrical calculations involving triangles and circles. Even the so-called imaginary number,[20] i ($\sqrt{-1}$), is used by electrical engineers to model the power of an alternating current (AC) circuit at any given time.[21]

In such a world, where the fantastical can be just as useful as the real, it can be difficult to distinguish truth from fantasy. It is therefore important to have some sort of mechanism for proving the truth value of mathematical propositions. The search for a solid foundation for mathematical truths is as much a question of the history of mathematics as a question of philosophy. And it is difficult terrain to traverse. There are nearly as many views on the nature of mathematics as there are mathematicians. No one school of thought has become dominant, despite considerable debate.

For present purposes, it is sufficient to look at a representative sample, as most alternative theories can be considered to be derivatives, mixtures, or variations of these basic theories. A final resolution of this debate within the pages of this book will not be attempted. An understanding of these theories is however central to this book, because the school of thought to which one subscribes greatly influences the reasoning by which it might be argued that mathematics is non-patentable. If it is asserted that mathematical objects and principles exist *a priori*, then mathematics is non-patentable because it amounts to a discovery. If one thinks they are merely mental constructions, then mathematics is non-patentable as abstract idea. If you take what might be called a inclusive view, that no one theory of mathematics will provide the ultimate guidance, but all have something useful to say, then the approach to reconciliation lies down a different path. That point will be returned to at the end of this chapter.

---

19 It is said that when the Greek mathematician Hipassus reported his discovery of the irrationality of $\sqrt{2}$ he "became so much hated by the Pythagoreans, that not only they cast him out of the community; they built a shrine for him as if he were dead". Hyamblicus of Chalkis, *De vita pythagoras*, at 246–247. Cited in Michael Lahanas, "Irrational numbers" <http://www.mlahanas.de/Greeks/Irrational.htm> (28 December 2007). Alternative accounts suggest he was thrown overboard from the ship he was on at the time and drowned. See Richard Mansfield, "Real and Complex Numbers" in *Logic: The Bridge to Higher Math* <http://www.math.psu.edu/melvin/logic/node8.html> (7 September 2011).

20 This derogatory term was coined by Descartes, who rejected their use because "while we can always conceive [of the square roots of negative numbers], still there is ... no quantity corresponding to those we conceive." See Rene Descartes, *Discourse on Method, Optics, Geometry, and Meteorology* (PJ Olscamp (trans), Hackett Publishing, 2001) at 236.

21 See for example, John A Masters, "Imaginary Numbers Really Exist! (I'm Really, Really Serious!)" on *Least Significant Bits* <http://jamspeaks.blogspot.com/2007/09/imaginary-numbers-really-exist-im.html> (7 September 2011).

To fully understand the philosophy of mathematics it is necessary to have some regard to the history of mathematics. A complete history of mathematics, in all its glorious detail, lies well beyond the scope of this book.[22] Instead, only a brief history is attempted, in order that some context might be given to the various philosophies of mathematics which have been put forward, and to tie the development of mathematics to the various milestones of the patent system.

1.1 The history and philosophies of mathematics

The origins of mathematics lies in the use of numbers, which on a conservative estimate predates the birth of the patent system by at least 3000 years.[23] At this earliest stage, "the crucial point that marks the emergence of mathematics from mere counting is .... that of abstraction."[24] But it was not until the ancient Greek civilisation that "mathematics would make the transition from a strictly practical discipline to an intellectual one, where knowledge for its own sake became the main goal."[25]

A Ancient times

For nearly 2000 years, mathematics was dominated by the Greeks,[26] whose "characteristic achievements [were] the development of rigorous deductive proofs and the geometrical representation of the cosmos for astronomical and geographical purposes."[27] In around 375 BC, Plato wrote *The Republic*, in which he posited a theory of knowledge in which the realm of the intelligible

---

22 In any event, such histories have already been compiled. See for example, Morris Kline, *Mathematical Thought from Ancient to Modern Times* (Oxford University Press, 1972); David Eugene Smith, *History of Mathematics* (Courier Dover Publications, 1958); Walter William Rouse Ball, *A Short Account of the History of Mathematics*, (4th ed, Project Gutenberg, 2010); Florian Cajori, *A History of Mathematics* (Project Gutenberg, 2010). For a more concise account see Dirk J. Struik, *A Concise History of Mathematics*, (3rd ed, Dover Publications, 1967); Robert Tubbs, *What is a Number? Mathematical Concepts and Their Origins* (John Hopkins University Press, 2009); Eli Maor, *To Infinity and Beyond: A Cultural History of the Infinite* (Birkhäuser, 1987).
23 The earliest source of information on the mathematics of the ancient Egyptians comes from the Moscow Mathematical Papyrus which dates from around 1850 B.C. See Anglin, *Mathematics, a concise history and philosophy* (Springer Verlag, 1994) at 1–2. Struik places the origins of mathematics at the time of the "transition ... from the mere *gathering* of food to its actual *production*, from hunting and fishing to agriculture.": Struik, above n22 at 7, which he puts at "perhaps ten thousand years ago". As noted above, the first recorded patent system was in Venice, in the 15th century.
24 McLeish, above n3.
25 Maor, above n22 at 31.
26 "To be sure, mathematics as a science had already reached quite an advanced stage before the Greek era... But the ancient mathematics of the Hindus, the Chinese, the Babylonians and the Egyptians confined itself solely to practical problems of daily life, such as the measurement of area, volume, weight and time": Maor, above n22 at 2–3.
27 Ian Mueller, 'Earlier Greek Mathematics' in Donald J. Zeyl, Daniel Devereux, and Phillip Mitsis (eds), *Encyclopedia of Classical Philosophy* (1997).

was to be given more weight than the realm of the visible.[28] The visible world was in a constant state of flux, and was capable of founding only opinion or belief. The "real" world and true basis for truth or knowledge was thus the intelligible world, wherein the immutable ideal forms (being archetypes or abstractions of visible-world objects) could be found similarly immutable truths.[29] As such, mathematics was both separate and superior to empirical sciences. Plato divided the subject matter of mathematics into number theory (arithmetic), geometry, astronomy, and harmonics (or music). These subjects made up the *Quadrivium*, the second stage of his seven liberal arts, which were to be mastered "before turning to dialectic and the ascent to the Good."[30] Within the *Quadrivium*, the subjects were paired, geometry with astronomy, and arithmetic with music, each pair corresponding to the abstract and concrete aspects of an area.[31]

In relation to practical mathematics, it was around 300 BC that Euclid systematized geometry[32] in *The Thirteen Books of the Elements*. Euclid's system was based

28 See Plato, "The Republic" in *The Dialogues of Plato translated into English with Analyses and Introductions by B. Jowett, M.A. in Five Volumes* (3rd ed, Oxford University Press, 1892) at line 511 (Stephanus numbering): "[c]orresponding to these four divisions [achieved by splitting the realms of the visible and intelligible into two], let there be four faculties in the soul—reason answering to the highest, understanding to the second, faith (or conviction) to the third, and perception of shadows to the last—and let there be a scale of them, and let us suppose that the several faculties have clearness in the same degree that their objects have truth."

29 "[T]he principal doctrines of the dialogues express Plato's own philosophy: There is a realm of eternal and changeless objects called "Ideas" or "Forms"; these are the most real entities and the basic objects of knowledge; the Form of the Good is the highest object of understanding; the observable world is a deficient reflection of the Forms...": Richard Kraut, "Plato (427–347 B.C.E.)" in Zeyl, above n27 at 390. In the *Phaedo*, Plato gives an example based on the notion of equality demonstrates the existence of the Forms. Kraut (at 397) summarises it as follows:

> Equal sticks are as much an example of equality as inequality, since each member of the pair is also unequal to other sticks; but the Form is precisely what equality is and can never be characterized as unequal. This is why someone who knows what equality is must have acquired this knowledge by looking to the Form and not merely to objects whose equality is detected by the senses.

30 Mueller, above n27 at 316, citing Plato's *Republic*.

31 Luigi Borzacchini, "2. Greek mathematics and Pythagoras" in *Being and Sign II: Axiomatic deductive method and infinite in Greek mathematics*, (1995) <http://www.dm.uniba.it/~psiche/bas2/node3.html> (2 November 2010).

32 It may be noticed that geometry seems to have preoccupied the ancient Greeks to a much greater extent than arithmetic. For example, it is suggested that geometry was "the most philosophically satisfying branch of mathematics by the ancient Greeks; and indeed until the end of the Middle Ages, the word 'mathematics' meant geometry": McLeish, above n2. Plato considered geometry to be "the clearest example of an human natural access to the Forms world": Borzacchini, above n31. This may in part be due to the limitations imposed on number theory by the Pythagoreans, whose theory of numbers extended only to integers (or whole numbers). As such, many problems involving non-integers (for example, so-called irrational numbers such as π, or $\sqrt{2}$) were "thus resolved ... in a geometric manner instead": Wilbur R Knorr, "Transcript of a Lecture Delivered at The Annual Convention of

upon five axioms, the truth of which was said to be self-evident. From these five axioms, "a vast amount of knowledge [was] derived in steps which are apparent to any reasonably intelligent and sufficiently well-trained reader."[33] However,

> its dependence on visual intuition (whose consequent deductive gaps were already noted by Archimedes), together with the challenge of Euclid's infamous fifth postulate (about parallel lines), and the famous unsolved problems of compass and straightedge construction, established an agenda for generations of mathematicians.[34]

Despite these limitations "[t]he *Elements* form, next to the Bible, probably the most reproduced and studied book in the history of the Western world."[35]

Platonism: Plato's theory of the Forms gives rise to the most long-lived of the philosophies of mathematics, platonism. On a platonist account, the realm of mathematics is populated with "abstract, necessarily existing objects, independent of the human mind."[36] These mathematical objects exist in the realm of the intelligible rather than the visible.[37] Despite the name, however, platonism is not limited to the terms of Plato's account, and is normally "defined and debated independently of its original historical inspiration."[38]

Platonist accounts support an objective notion of proof which puts mathematics on a similar basis to physics. As a consequence, "[t]he statements of mathematics are true or false depending on the properties of those [abstract mathematical] entities, independent of our ability, or lack thereof, to determine which."[39] It is perhaps for this reason that Platonist accounts continue to enjoy a popularity with mathematicians, at least as a starting point,[40] because the

> Platonist attitude to objects of investigation is inevitable for a mathematician: during his everyday work he is used to treat numbers, points, lines

---

the History of Science Society, Atlanta, December 28, 1974" in Jean Christianidis (ed) *Classics in the history of Greek mathematics* (Kluwer Academic Publishers, 2004).

33 McLeish, "Mathematics" above n2.

34 CJP, "Philosophy of Mathematics" in Robert Audi (ed), *The Cambridge Dictionary of Philosophy* (2nd ed, Cambridge University Press, 1999) at 681.

35 Struik, above n22 at 50.

36 Anglin, above n23 at 218.

37 See page 7 above.

38 Øystein Linnebo, "Platonism in the Philosophy of Mathematics," *Stanford Encyclopedia of Philosophy*, 18 July 2009 <http://plato.stanford.edu/entries/platonism-mathematics/> (22 November 2010).

39 Penelope Maddy, *Realism in Mathematics* (Clarendon Press, 1990) at 21.

40 "In everyday life, we speak as Platonists, treating the objects of our study as real things that exist independently of human thought. If challenged on this, however, we retreat to some sort of formalism, arguing that in fact we are just pushing symbols around without making any metaphysical claims. Most of all, however, we want to do mathematics rather than argue about what it actually is. We're content to leave that to the philosophers." Jean Dieudonné. Cited in Fernando Gouvea, "Book Review: What is Mathematics Really? by Reuben Hersh," *MAA Online*, <http://test.maa.org/reviews/whatis.html> (9 September 2011).

etc. as the 'last reality', as a specific independent 'world'. This sort of platonism is an essential aspect of mathematical method, the source of the surprising efficiency of mathematics in the natural sciences and technology.[41]

Indeed some "working realists" seek to avoid the philosophical debate entirely, by asserting a "methodological view that mathematics should be practiced *as if* platonism was true".[42]

This other-worldliness raises questions, however, namely "[w]here is this world and how do we make contact with it? How is it possible for our mind to have an interaction with the Platonic realm so that our brain state is altered by that experience?"[43] In other words, accepting the Platonist view of mathematical entities requires a leap of faith.[44]

It is perhaps for this reason that modern interpretations of the Platonist view abandon the epistemological claims as to how to access the world of Forms might occur. Brown for example points out that ordinary perception of the physical world may be understood as a matter of physiology, but how those perceptions become sensations and beliefs "is a very great mystery. It is just as great a mystery as how mathematical entities bring about mathematical beliefs."[45] But for the discipline of mathematics, with its "virtually unparalleled standard of intellectual rigor and exactitude",[46] this explanation remains a little unsatisfying.

B   15th century

The first recorded grant of a patent was set in Venice in the 15th century, with many other European states having granted similar rights in the 15th and 16th centuries.[47] At this time, the theoretical province of mathematics was still

---

41 Karlis Podnieks, "Platonism, Intuitionism and the Philosophy of Mathematics" in (1992) 31 *Semiotika i informatika* 150, English translation available at <http://www.ltn.lv/~pod­nieks/gt1.html> (29 Oct 2007).

42 Linnebo, above n38. Linnebo cites Bernays and Shapiro as adherents of this view: Paul Bernays, "On Platonism in Mathematics" in Paul Benecerraf and Hilary Putnam (eds) *Philosophy of Mathematics: Selected Readings* (2nd ed, Cambridge University Press, 1983); Stewart Shapiro, *Philosophy of Mathematics: Structure and Ontology* (Oxford University Press, 1997) at 21–27 and 38–44.

43 John D Barrow, *Pi in the Sky: Counting, Thinking, and Being* (Clarendon Press, 1992) at 272.

44 "Change "mathematics" to "God" and little else might seem to change. The problem of human contact with some spiritual realm, of timelessness, of our inability to capture all with language and symbol – all have their counterparts in the quest for the nature of Platonic mathematics.": Barrow, above n43 at 272.

45 James Robert Brown, *Philosophy of Mathematics: An Introduction to the World of Proofs and Pictures* (Routledge, 1999) at 15.

46 Michael J White, "Plato and Mathematics," in Hugh H Benson (ed), *A Companion to Plato* (Blackwell Publishing, 2005) at 228.

47 See See Edward C Walterscheid, "The Early Evolution of the United States Patent Law: Antecedents (Part 1)" (1994) Journal of the Patent and Trademark Office Society 697 at 708 (footnote 37).

dominated by the *Quadrivium*, which was taught at medieval universities.[48] Over this period, the "Italians, and particularly the Venetians, realized the importance of the use of arithmetic in their daily business transactions",[49] giving rise to innovations such as double-entry bookkeeping,[50] and incremental advances in computational methods and improved notation systems, which would eventually lead to the emergence of algebra in its modern form.[51]

Also at this time, the development of geometry was being driven by Renaissance artists in a number of ways "for translating the reality of 3-dimensional natural phenomena onto 2-dimensional surfaces, producing virtually realistic copies"[52] (perspective), and also "in determining the correct proportions for the figures they drew"[53] (proportion). Leonardo da Vinci is perhaps the best known exponent, but is by no means the only example.[54]

Also contributing to the development of mathematics over this period was "the productive use and further perfection of machines"[55] which led to "theoretical mechanics and the scientific study of motion and of change in general".[56]

C 17th century

The traditional limitations on mathematics, requiring a close correlation with the real world, began to be questioned in the 17th century when mathematicians such as Descartes began to approach geometrical problems using the tools of algebra.[57] When the *Statute of Monopolies* was passed at the beginning of the

---

48 "The early and central Middle Ages had inherited from antiquity the scheme of the seven Liberal Arts: grammar, rhetoric, dialectics (... the *trivium*), and the four mathematical arts, arithmetic, geometry, music and astronomy (the *quadrivium*).": Jens Høyrup, *In measure, number, and weight: studies in mathematics and culture* (SUNY Press, 1994) at 177. "The quadrivium functioned as the source of theoretical and exact science for medieval university students.": Edward Grant, *The foundations of modern science in the Middle Ages: their religious, institutional and intellectual contexts* (1996) at 44. See also Frank Swetz, *Capitalism and Arithmetic: The New Math of the 15th Century* (David E Smith (trans), Open Court Publishing, 1987) at 14, noting the lack of prominence given to arithmetic.

49 Swetz, above n48 at 11.

50 Swetz, above n48 at 12.

51 Struik, above n22 at 93–97.

52 Joseph W Dauben, "Galileo and Perspective: The Art of Renaissance Science" *American Physical Society*, 2002, <http://www.aps.org/publications/apsnews/200201/backpage.cfm> (21 December 2007).

53 Dauben, above n53.

54 See Robert Tubbs, *What is a Number? Mathematical Concepts and Their Origins* (John Hopkins University Press, 2009) at 141–151 who notes Alhazen's theory of vision, as well as the work of di Bondone and Brunelleschi, in addition to da Vinci. See also Struik, above n22 at 90.

55 Struik, above n22 at 99.

56 Struik, above n22 at 99.

57 Helena M Pycior, "Mathematics and Philosophy: Wallis, Hobbes, Barrow, and Berkeley" (1987) 48(2) *Journal of the History of Ideas* 265 at 267. See also Struik, above n22 at 102–104.

17th century, mathematics was in a state of flux. This period was dominated by issues such as "the relative merits of geometry versus arithmetic and algebra centred on questions of the ability of the human mind to reason on symbols, the legitimacy of the negative and imaginary numbers, and finally the ultimate source of the numbers of arithmetic."[58] These illustrate the concern amongst mathematicians and philosophers about the increasingly abstract nature of the subject matter.

Debates as to the nature of mathematics at this time were dominated by two schools of thought, namely rationalism and empiricism. Rationalism was supported largely by algebraists such as Descartes, Spinoza, and Liebniz. Empiricism was championed by (largely English) geometers, such as Locke, Berkeley, Hume, and J.S. Mill.

Rationalism: The rationalist account is similar in many respects to platonism,[59] although it does not assume the independent existence of mathematical objects.[60] Rationalists posit that both the source and ultimate arbiter of knowledge is human reason. For example, Descartes suggested that reason "is a more powerful instrument of knowledge than any other that has been bequeathed to us by human agency, as being the source of all others".[61]

Empiricism: Empiricism asserts that mathematics ultimately depends for its validation on evidence that is observable by the senses. That is, mathematical truths are verified by empirical research, just as in any other science. This approach is underscored by a

> view that the world is largely, or even, entirely a product of chance. On the empiricist account, the universe consists of many independent individuals, which, if they are connected, are so only accidentally, reducing causation to nothing more than a matter of constant conjunction. ... Under

---

58 Pycior, above n57 at 267.

59 This is so in that rationalist accounts maintain a distinction between the visible and intelligible realms, and place an emphasis on the intelligible as superior.

60 "[A] signature doctrine of rationalism is the doctrine of innate ideas, according to which the mind has built into it not just the structure of knowledge but even its content": Thomas M Lennon and Shannon Dea, "Continental Rationalism" *Stanford Encyclopedia of Philosophy*, 21 November 2007 <http://plato.stanford.edu/entries/continental-rationalism/> (27 Sep 2010). However, note that Descartes' theory as to the nature of mathematical objects (René Descartes, "Meditation V: On the Essence of Material Objects and More on God's Existence" in *Meditations on First Philosophy* (Areté Press, 1986) at 35):

> Suppose, for example, that I have a mental image of a triangle. While it may be that no figure of this sort does exist or ever has existed outside my thought, *the figure has a fixed nature (essence or form), immutable and eternal, which hasn't been produced by me and isn't dependent of my mind.* (emphasis added)

61 René Descartes, "Rules for the Direction of the Mind" in *Key Philosophical Writings* (Wordsworth Editions, 1997) at 13.

such circumstances, only experience of the world can provide knowledge of it.[62]

John Locke,[63] George Berkeley,[64] and David Hume[65] were the primary exponents of empiricism.[66] J.S. Mill was also an important early supporter of the notion of mathematics as an empirical science.[67] Mill argued for example that the laws of arithmetic were nothing more than "inductive generalisations from observed facts".[68]

62 Lennon and Dea, above n60.
63 John Locke, *An Essay Concerning Human Understanding*, (John W. Yolton (ed), Dent, 1976). According to Locke, the mind at birth "is a *tabula rasa* or blank sheet until experience in the form of sensation and reflection provide the basic materials – simple ideas – out of which most of our more complex knowledge is constructed.": William Uzgalis, "John Locke," *Stanford Encyclopedia of Philosophy*, 2 September 2001, <http://plato.stanford.edu/entries/locke/> (30 November 2010). For Locke, mathematics was a "true method of advancing knowledge ... by considering our abstract ideas... [M]athematicians, ... from very plain and easy beginnings, by gentle degrees, and a continued chain of reasonings, proceed to the discovery and demonstration of truths that appear at first sight beyond human capacity.": Locke, above, Book IV, Chapter 12, paragraph 7.
64 George Berkeley, *Treatise Concerning the Principles of Human Knowledge* (Project Gutenberg, 2009) <http://www.gutenberg.org/files/4723/4723-h/4723-h.htm> (29 November 2010). Unlike Locke, Berkeley attributed to mathematics a merely secondary role, noting that "how celebrated soever [mathematics] may be for [its] clearness and certainty of demonstration, which is hardly anywhere else to be found, [it] cannot nevertheless be supposed altogether free from mistakes": at [108].
65 David Hume, *An Enquiry Concerning Human Understanding* (2006, Project Gutenberg) <http://www.gutenberg.org/dirs/etext06/8echu10h.htm> (29 November 2010). Similarly to Berkeley, Hume seems satisfied with mathematics to the extent that it maintains a connection with the physical world, but notes that "the finer sentiments of the mind, the operations of the understanding, the various agitations of the passions, though really in themselves distinct, easily escape us, when surveyed by reflection; nor is it in our power to recall the original object, as often as we have occasion to contemplate it. Ambiguity, by this means, is gradually introduced into our reasonings: Similar objects are readily taken to be the same: And the conclusion becomes at last very wide of the premises": at [48].
66 Mention should be made at this point of Aristotle, whose philosophy "contains seeds of empiricism": Stewart Shapiro, *Thinking About Mathematics* (Oxford University Press, 2000) at 62. Aristotle agreed with Plato as to the existence of Forms (or universals, as he called them). However, Aristotle contended that these forms "exist in perceptible objects": Aristotle, *Metaphysics*, Book M, (Annas (trans), Clarendon Press, 1976), cited in Shapiro at 64. These perfect forms are accessed through a faculty of abstraction, in which physical objects are contemplated, and some of their features are abstracted away. See Shapiro at 66.
67 See John Stuart Mill, *A System of Logic Ratiocinative and Inductive being a Connected View of the Principles of Evidence and the Methods of Scientific Investigation*, (8th ed, Harper & Brothers, 1882) <http://www.gutenberg.org/files/27942/27942-pdf.pdf> (10 September 2011), and in particular Chapters 5 and 6 of Book II, Chapter 24 of Book III. For a summary of Mill's position, see Donald Gillies, "An Empiricist Philosophy of Mathematics and Its Implications for the History of Mathematics," in Emily Grosholz and Herbert Breger (eds), *The Growth of Mathematical Knowledge* (Springer, 2000) at 41–57.
68 Gillies, above n67 at 41.

What both empiricist and rationalist accounts have in common is a harmony between mathematics and the real world, and in particular, physics. Whilst the Greeks had maintained a sharp distinction between it and science, the mathematics of the time of Descartes and Newton was not so distinct from physics. Kline notes that

> as the province of mathematics expanded, and mathematicians not only relied upon physical meanings to understand their concept but accepted mathematical arguments because they gave sound physical results, the boundary between mathematics and science became blurred.
>
> ... In fact, it would be difficult to name an outstanding mathematician of the [18th] century who did not take a keen interest in science. As a consequence these men did not wish or seek to make any distinctions between the two fields.[69]

Transcendental idealism: Kant's transcendental idealism stands halfway between rationalist and empiricist accounts.[70] Rather than the typical empiricist account by which sensory perceptions were passively received, Kant held that "the mind of the knower makes an active contribution to experience of objects before us."[71] For Kant, "[a]ppearances are not things in themselves. Empirical intuition is possible only through the pure intuition (of space and time)."[72] It is by reference to this same "pure intuition" that Kant believed mathematical concepts were constructed, giving them the same verifiability as sensory information, and thereby differentiating mathematics from other intellectual pursuits such as philosophy. Kant's theory was "developed ... in the context of the actual mathematical practices of his predecessors and contemporaries, and he produced thereby a coherent and compelling account of early modern mathematics."[73] As a result, Kant's theory was to dominate until the upheaval of mathematics in the second half of the 19th century.

Whilst the rationalist/empiricist views stand in stark contrast, they do have a common aspect. Both seem to accept that the doing of mathematics is an activity set apart from the physical world. On either view, mathematics is a mental activity which is true by virtue of its reliance on reason, or verified after the fact through empirical observation of the physical world.

---

69 Kline, above n22 at 395.

70 What follows is a broad brush account of Kant's philosophy of mathematics. For a more detailed account see Lisa Shabel, "Kant's Philosophy of Mathematics" in Paul Guyer (ed) *The Cambridge Companion to Kant and Modern Philosophy* (Cambridge University Press, 2006).

71 Matt McCormick, "Immanuel Kant: Metaphysics" *Internet Encyclopedia of Philosophy*, 30 June 2005 <http://www.iep.utm.edu/kantmeta/> (7 December 2010).

72 Immanuel Kant, *Critique of Pure Reason* (Paul Guyer, Allen W Woods (trans, eds), Cambridge University Press, 1998) at 289 (A165/B206).

73 Shabel, above n70 at 119.

D   19th century

By the 19th century, the rationalist/empiricist debate had been largely resolved in favour of empiricism.[74] By the second half of the century, however, the growing abstractness and complexity of mathematics was causing the link between the external world and the study of mathematics to be questioned.

In particular, Mill's views were harshly satirised by Gottlob Frege, the founder of the logicist school,[75] who remarked:

> What, then, are we to say of those who, instead of advancing this work where it is not yet completed, despise it, and betake themselves to the nursery, ... there to discover, like John Stuart Mill, some gingerbread or pebble arithmetic![76]

Frege pointed to our understanding of operations on large numbers to illustrate the limitations of Mill's philosophy. He pointed out that "[o]n Mill's view we could actually not put $1,000,000 = 999,999 + 1$ unless we had observed a collection of things split up in precisely this peculiar way."[77]

Over the nineteenth century, "as in the two preceding centuries ... the progress in mathematics brought with it larger changes barely perceptible in the year-to-year developments but vital in themselves and in their effect on future developments."[78] Because of the "vast expansion in subject matter and in the opening of new fields as well as the extension of older ones",[79] "by 1870 mathematics had grown into an enormous and unwieldy structure, divided into a large number of fields in which only specialists knew the way."[80] A desire to synthesize these various fields lead to the a new set of unifying principles centred around group theory, which has been described as "a supreme example of the art of mathematical abstraction".[81]

In this period, Gauss, Lobachevsky, Bolyai, and Riemann challenged the supremacy of Euclidean geometry by demonstrating that it was possible to construct an alternative geometry in which Euclid's fifth postulate did not hold.[82] The development of non-Euclidean geometries was "from the standpoint of intellectual importance and ultimate effect on the nature of mathematics, the most

---

74 Leon Horsten, "Philosophy of Mathematics" *Stanford Encyclopedia of Philosophy*, 25 September 2007 <http://plato.stanford.edu/entries/philosophy-mathematics> (10 September 2011)
75 The logicist philosophy of mathematics is considered below.
76 Gottlob Frege, *The Basic Laws of Arithmetic* (Montgomery Furth (trans), University of California, 1967) at vii.
77 Frege, above n76 at 10–11.
78 Kline, above n22 at 1023.
79 Kline, above n22 at 1023.
80 Struik, above n22 at 1024.
81 James R. Newman, *The World of Mathematics, Volume 3* (Courier Dove Publications, 2000) at 1534.
82 Struik, above n22 at 167.

consequential development [in mathematics]".[83] That such geometries could be internally consistent suggested that

> the set of axioms one chooses as the foundation of a mathematical structure [such as geometry] is, to a certain extent, arbitrary; change one or more of the axioms, and a different structure will emerge. Whether the new structure agrees with the 'real' physical world is totally irrelevant; what matters is logical consistency alone.[84]

Geometry then collapsed "into disfavour because mathematicians found that they had unconsciously accepted facts on an intuitive basis, and their supposed proofs were consequentially incomplete. The danger that this would continually recur made them believe that the only sound basis for geometry would be arithmetic."[85] However, "even arithmetic and the analysis built on it soon became suspect. The creation of non-commutative algebras ... raised the question of how one can be sure that ordinary numbers possess the privileged property of truth about the real world."[86] The result was that "[b]y 1900 mathematics had broken away from reality; it had clearly and irretrievably lost its claim to the truth about nature, and had become the pursuit of necessary consequences of arbitrary axioms about meaningless things."[87]

The extent of the relationship between mathematics and the "real" world continued to be a subject of intense debate. Some, such as Cantor, saw the freedom from reality as that which distinguishes mathematics from other fields:

> Mathematics is entirely free in its development and its concepts are restricted only by the necessity of being noncontradictory and coordinated to concepts previously introduced by precise definitions. ... The essence of mathematics lies in its freedom.[88]

Others lamented "[t]he loss of truth and the seeming arbitrariness, the subjective nature of mathematical ideas and results."[89] Felix Klein, for example, himself a major figure in the development of non-Euclidean geometries, considered the investigation of such arbitrary structures as "the death of science", and felt

---

83 Kline, above n22 at 1023. The impact of non-Euclidean geometries was however greatly magnified by the introduction of more and more concepts with little physical relevance into the domain of mathematics, from irrational to negative and complex numbers, "quaternions, ... complex elements in geometry, *n*-dimensional geometry, bizarre functions, and transfinite numbers": at 1029.

84 Maor, above n22 at 124.

85 Kline, above n22 at 1016.

86 Kline, above n22 at 1034.

87 Kline, above n22 at 1035.

88 Georg Cantor, "Über unendliche lineare Punktmannigfaltigkeiten" (1883) *Math. Ann* 21, cited in Kline, above n22 at 1031. See also Jacobi, *Ges Werke*, 1, 454–55, cited in Kline at 1037.

89 Kline, above n22 at 1035.

that "whoever has the privilege of freedom should also bear responsibility [of using it to investigate nature]."[90] Similarly, "Hilbert not only stressed that concrete problems are the lifeblood of mathematics, but took the trouble in 1900 to publish a list of twenty-three outstanding ones."[91] Understandably, these differing viewpoints left an enduring division in the field between so-called pure and applied mathematicians.[92]

Platonism remained popular during this period, since on a platonist account, non-Euclidean geometries exist, no less than Euclidean geometries. Correspondence with our experience of the physical world is not required. However, a number of new attempts were made to shore up the foundations of mathematics, and now fall to be discussed.

Logicism: The central tenet of logicism is that all of mathematics is merely a branch of logic. As such it is a form of rationalism. The logicist school was founded by Gottlob Frege, who hoped to show "that arithmetic is a branch of logic and need not borrow any ground of proof whatever from either experience or intuition".[93] This "required that he show that the latter are derivable using only rules of inference, axioms, and definitions that are purely analytic principles of logic".[94] Frege's work on grounding mathematics in logic was almost complete when Bertrand Russell discovered a contradiction within its exposition of set theory,[95] an obstacle which Frege never surmounted.

---

90 Kline, above n22 at 1037.

91 Kline, above n22 at 1038.

92 Kline, above n22 at 1036. The division is complicated however by the way in which subjectively "pure" mathematics has underscored cutting-edge applications. For example, non-Euclidean geometries were key to Einstein's development of his theory of relativity, since the theory requires that

> space-time must be curved ... To describe such a four-dimensional curved space mathematically, the young physicist was looking for some kind of non-Euclidean geometry, and he found it in Riemann's geometry, which allows for space to have a variable curvature. ... General relativity, therefore, can be said to be the final triumph of a mathematical idea which, in its infancy, was no more than an intellectual exercise.

93 Frege, above n76 at §0, 29.

94 Edward N. Zalta, "Frege's Logic, Theorem and Foundations for Arithmetic," *Stanford Encyclopaedia of Philosophy*, 13 April 2007, <http://plato.stanford.edu/entries/frege-logic/> (27 March 2010) at 6.1.

95 The paradox, known as Russell's paradox, arose from the ability to define within Frege's system a set of all the sets which were not members of themselves. Kline sources the contradiction not in Frege's system, but in set theory itself. See Kline, above n22 at 1183. Kline also sets out a popular form of the paradox as follows:

> A village barber, boasting that he has no competition, advertises that of course he does not shave those people who shave themselves, but does shave all those who do not shave themselves. One day it occurs to him to ask whether he should shave himself. If he should shave himself, then by the first half of his assertion, he should not shave himself; but if he does not shave himself, then in accordance with his boast, he must shave himself. The barber is in a logical predicament.

Bertrand Russell and Alfred North Whitehead made their own attempt at grounding mathematics in logic, *Principia Mathematica*. They used formalist methods[96] that all statements of number theory could be derived from logical axioms. Russell and Whitehead claimed to base "all of mathematics on a logical system derivable from five primitive logical propositions whose truth is founded on basic intuition".[97] Russell and Whitehead's attempt was ultimately undone by Kurt Gödel, whose contribution is discussed below.[98]

Formalism: Formalists, led by David Hilbert, focused on expressing mathematics as formal logical systems and studying them without considering their meaning. To formalists, mathematics is "no more or less than mathematical language. It is simply a series of games".[99] Mathematicians may read meanings into particular terms, but the terms themselves are without meaning. The truth value of mathematical theorems to a formalist depends on their provability using the rules of the system. The success of any particular formal system depends on three things: consistency, completeness, and decidability.[100]

A particularly painful thorn in the formalists' side came in the form of a German mathematician, Kurt Gödel, who in proving Russell and Whitehead's project flawed, came up with a mathematical proof called the incompleteness theorem, which can be summarised as follows:

> Any sufficiently powerful formal system cannot be both complete and consistent.[101]

Although Gödel's proof only strictly extends to mathematical systems, Nagel and Newman suggest the theorem applies to "a very large class of deductive systems".[102] In any event, Gödel's theorem dealt a fatal blow to the formalist program, at least as a foundation for mathematics.

Interestingly though, the formalist notion of reducing the mental processes of 'doing mathematics' as a series of finite, mechanical steps bore unexpected fruit.

---

96   See the discussion of formalism below.

97   Berlinghoff, above n6 at 203.

98   Cf. Bernard Linsky and Edward N. Zalta, "What is Neologicism?" (2006) 12(1) *The Bulletin of Symbolic Logic* 60 at 61: "As we look back at logicism, we shall see that its failure is no longer such a clear-cut matter…". The authors go on to argue a new breed of neologicism which "closely approximates the main goals of the original logicist programme."

99   Anglin, above n23 at 218.

100 "Consistency means that no contradictions will be found in the system, i.e., a theorem and its negation cannot both be true. Completeness implies that every theorem which is true in the system can be proved within the system. Decidability requires that a finite mechanical procedure exists that determines whether any given claim made by the system may be proved within the system.": Brigham Narins (ed), *World of Mathematics*, Volume 1 (Gale Group, 2001) at 235.

101 The above simplification of the theorem is in this author's words, but is is based on a number of different formulations. See Douglas R. Hofstadter, *Godel, Escher Bach: An Eternal Golden Braid* (Basic Books, 1979); Narins, above n100 at 235.

102 Ernest Nagel and James R. Newman, *Gödel's proof* (NYU Press, 2001) at 5.

In 1936, Alan Turing developed "a formal counterpart to the notion of a mental process",[103] the Turing machine, which became the theoretical model for the digital computer. So it seems that incompleteness and inconsistency are no insurmountable obstacle to the development of software, as those who have done battle with a computer can surely attest.[104]

Intuitionism: Intuitionists, like L.E.J. Brouwer and Heyting, suggest that mathematics is "a creation of the human mind. Numbers, like fairy tale characters, are merely mental entities which would not exist if there were never any human minds to think about them."[105] For intuitionists, the truth of mathematical propositions is not discovered, it must be experienced.[106] As a foundation for mathematics, intuitionists start from the idea that "[t]he basic notions of mathematics are so extremely simple, even trivial, that doubts about their properties do not rise at all."[107] Heyting began with an intuitive understanding of counting, which involves the isolation of one object after another. Since "[i]solating an object, focusing our attention on it, is a fundamental function of our mind," he derives the notion that "[t]he entity conceived in the human mind is the starting point of all thinking, and in particular of mathematics."[108] Heyting then goes on to construct arithmetic, and then to deal with more complex mathematical systems such as the continuum and set theory.[109]

One interesting aspect of the intuitionist position is that "[i]n its simplest form mathematics remains confined to one mind."[110] This means that mathematical formulae, which for the formalist are the ultimate expression of mathematics, are merely the language of mathematics. As with any language then, they are

---

103 Jean Lassègue, "Doing Justice to the Imitation Game; a Farewell to Formalism," December 2003, <http://formes-symboliques.org/article.php3?id_article=75> (10 September 2011).

104 The relationship between Gödel's theorem, and Turing's model is the theoretical aspect of the isomorphism between mathematics and software development, which was touched on in Chapter 1 in Section 3 on page 14, and is explored further in Chapter 5 in Section 2 on page 166.

105 Anglin, above n23 at 219.

106 "[O]n Brouwer's view, there is no determinant of mathematical truth outside the activity of thinking, a proposition only becomes true when the subject has experienced its truth (by having carried out an appropriate mental construction); similarly, a proposition only becomes false when the subject has experienced its falsehood (by realizing that an appropriate mental construction is not possible). Hence Brouwer can claim that 'there are no non-experienced truths' ([L.E.J. Brouwer, *Collected Works*, Volume 1, Philosophy and Foundations of Mathematics (A Heyting (ed), North-Holland, 1975)], p. 488).": Mark van Atten, "Luitzen Egbertus Jan Brouwer" in Edward N. Zalta (ed), *Stanford Encyclopaedia of Philosophy*, (Summer 2011 Edition) <http://plato.stanford.edu/archives/sum2011/entries/brouwer/> (10 September 2011).

107 Arend Heyting, "Intuitionistic Views of the Nature of Mathematics" (1974) 27 *Synthese* 79 at 79.

108 Heyting, above n107 at 80.

109 Heyting, above n107 at 81–88.

110 Heyting, above n107 at 80.

"not immune from misunderstanding"[111] since "mental constructions cannot be rendered exactly by means of language."[112]

The biggest criticism of the intuitionist position is a pragmatic one. Intuitionists necessarily reject concepts like the infinite[113] and the law of the excluded middle.[114] This makes it much harder to actually *do* mathematics.[115]

Summary: As the 19th century drew to a close, these various schools were in a standoff. No one theory had become dominant, and none appeared likely to become so. In time, the debate was simply set aside, so mathematicians could get on with *doing* mathematics.

E   *20th century*

In a sense, the development of mathematics over the 20th century continued the trends of the 19th century. Consistent with the move from Euclidean geometries to non-Euclidean geometries in physics (Newton to Einstein), "the emphasis [of mathematics] has shifted [from studying things on a small scale, in local coordinates] to try and understand ... global, large-scale behaviour."[116] Other progressions involved the study of increased dimensions,[117] the shift from studying commutative to non-commutative systems,[118] from linear to non-linear

---

111 Heyting, above n107 at 88.
112 Heyting, above n107 at 89. See also Kline, above n22 at 1200–1201:

> The world of mathematical intuition is opposed to the world of causal perceptions. In this causal world, not in mathematics, belongs language, which serves there for the understanding of common dealings. Words or verbal communications are used to communicate truths. Language serves to evoke copies of ideas in men's minds by symbols and sounds. But thoughts can never be completely symbolized. ... Mathematical ideas are independent of the dress of language and are in fact much richer.

113 The intuitionist position with respect to the infinite is perhaps more subtle than might be initially thought. Kline notes that "For Brouwer, as for all intuitionists, the infinite exists in the sense that one can always find a finite set larger than the given one. To discuss any other type of infinite, the intuitionists demand that one give a method of constructing or defining this infinite in a finite number of steps.": Kline, above n22 at 1202–1203.
114 This rule states that all things are either true or false. By excluding this rule, it is not possible for an intuitionist to prove the truth of a proposition by showing that it is not false.
115 Kline noted in 1972 that "[t]hey have succeeded in saving the calculus with its limit processes, but their construction is very complicated. They also reconstructed elementary portions of algebra and geometry. Unlike Kronecker, Weyl and Brouwer do allow some kinds of irrational numbers.": Kline, above n22 at 1203.
116 Michael F. Atiyah, "The Evolution of Mathematics in the 20th Century," (2001) 108(7) *The American Mathematical Monthly* 654 at 654.
117 In geometry, this meant a move beyond "things you could really see in space [to h]igher dimensions [which are] slightly fictitious, things that you could imagine mathematically": Atiyah, above n116 at 656. In algebra, which was "always concerned with more variables" this meant moving "from finite dimensions to infinite dimensions, from linear space to Hilbert space, with an infinite number of variables": Atiyah at 656.
118 Such as matrices and quaternions. A matrix is "A rectangular array of symbols or mathematical expressions arranged in rows and columns, treated as a single entity": "matrix" in *Oxford English Dictionary Online*, June 2011, <http://oed.com/view/Entry/

systems.[119] These developments evidence the further move of mathematics towards abstractions much more complicated than anything which might form the subject of human experience of a three-dimensional, linear world. A necessary consequence of this complexity was that the first half of the 20th century saw an increase in specialisation, as the foundational crisis in mathematics lead to its rigorous formalisation.[120] This in itself provides a challenge to the notion of proof, given that much of modern mathematics is beyond the expertise of all but the few mathematicians working in that area.[121]

In contrast to the earlier half of the 20th century, the latter half "is much more … the 'era of unification', where borders are crossed, techniques have been moved from one field into the other, and things have become hybridized to an enormous extent."[122] Finally, it must be noted the extent to which mathematics has been, at least since the last quarter of the 20th century, influenced by developments in physics. Atiyah noted in 2001 that

> [i]n the last quarter of the 20th Century, the one we have just been finishing, there has been a tremendous incursion of new ideas from physics into mathematics. … The results predicted by the physicists have time and again been checked by the mathematicians and found to be fundamentally correct, even though it is quite hard to produce proofs and many of them have not yet been fully proved.[123]

115057> (10 September 2011). A quaternion, also known is a "four-dimensional hypercomplex number that consists of a real dimension and 3 imaginary ones (i, j, k) that are each a square root of –1. They are commonly used in vector mathematics and three-dimensional games.": "Quaternion" in *Wiktionary*, 19 October 2010 <http://en.wiktionary.org/wiki/quaternion> (17 Nov 2010). All the laws of algebra apply to quaternions, except multiplication, which is non-commutative.

119 Non-Euclidean geometries, discussed above, are an example of a non-linear system of geometry.

120 The formalist approach to mathematics, discussed in further detail below, avoids the dependence on axioms as "'self-evident truths' … in favour of an emphasis on such logical concepts as consistency and completeness": "20th Century Mathematics" *The Story of Mathematics*, <http://www.storyofmathematics.com/20th.html> (18 November 2010). See also Atiyah, above n116 at 665.

121 See Henk Barendregt and Freek Wiedijk, "The Challenge of Computer Mathematics" (2005) 363 *Philosophical Transactions: Mathematical, Physical and Engineering Sciences* 2351 at 2352: "During the course of history of mathematics [*sic*] proofs increased in complexity. In particular, in the 19th century, some proofs could no longer be followed easily by just any other capable mathematician: one had to be a specialist. This started what has been called the sociological validation of proofs [by peer review]. … In the 20th century, this development went to an extreme." The authors observe the proof of the Classification of the Finite Simple Groups in 1979 which "consisted of a collection of connected results written down in various places, totalling 10 000 page … [and] also 'well-known' results [some of which turned out not to be valid]": *ibid* at 2352. It was 2004 before the proof was finally settled.

122 Atiyah, above n116 at 665.

123 Atiyah, above n116 at 663.

In the midst of all of this stands one particularly important development, the emergence of the theory of computability, and the development of the computer. In addition to having a use outside the field of mathematics, computers have come to be a useful tool in dealing with the increased complexity of mathematics, with computers being used to "prove" mathematical theorems which would otherwise involve too much computation.[124] Computers have also provided a source of data for feeding mathematical intuitions in circumstances where exact proofs are difficult or impossible.[125]

In light of this hybridisation and increasing complexity of subject matter, it should come as no surprise that the corresponding developments in the philosophy of mathematics became more subtle, and combined features of earlier schools of thought.

Quasi-empiricism: Quine and Putnam's *indispensability argument* claimed to address the reliance of Platonism on faith alone by arguing "that the mathematics that is used in physical theories is confirmed along with those theories and that scientific realism entails mathematical realism".[126] As such, this theory can be regarded as a mixture of empiricism and Platonism, hence the name quasi-empiricism. The Quine-Putnam argument can be summarised as follows:

Proposition 1: We ought to have ontological commitment to all and only the entities that are indispensable to our best scientific theories.

Proposition 2: Mathematical entities are indispensable to our best scientific theories.

Conclusion: We ought to have ontological commitment to mathematical entities.[127]

By implication, this limits the justification of mathematics to "enough mathematics to serve the needs of science".[128]

---

124 For example the Four-Colour problem and Kepler's Conjecture. See the discussion in Barendregt and Wiedijk, above n121 at 2352. For a discussion of the four-colour problem see Tubbs, above n22 at 269–270. See Brown, above n45 at 154–158 for a discussion of the status of such a proof as a "new way of doing mathematics ... [which is] *not* a priori, [is] *not* certain, [is] *not* surveyable, and [is] *not* open to double-checking by other mathematicians".

125 See Brown, above n45 at 158–171, wherein the author discusses conjectures and open problems in mathematics such as the qualities of "perfect" numbers, whether p is "normal", and the Riemann hypothesis.

126 Susan Vineberg, "Confirmation and the Indispensability of Mathematics to Science" (1996) 63(3) *Philosophy of Science* S256 at S256.

127 Adapted from Mark Colyvan, "Indispensability Arguments in the Philosophy of Mathematics" in Edward N. Zalta (ed) *The Stanford Encyclopedia of Philosophy* (Spring 2011 Edition) <http://plato.stanford.edu/archives/spr2011/entries/mathphil-indis/> (28 June 2011).

128 Colyvan, above n127.

The weak point of the indispensability argument lies within the first proposition.[129] Quine and Putnam found support for this proposition in a combination of naturalism and holism. Naturalism is the philosophical doctrine that "philosophy is neither prior to nor privileged over science,"[130] based on a "deep respect for scientific methodology and an acknowledgment of the undeniable success of this methodology as a way of answering fundamental questions about all nature of things."[131] However, naturalism "may or may not tell you whether to believe in all the entities of your best scientific theories",[132] so holism[133] completes the picture by positing that "our statements about the external world face the tribunal of sense experience not individually but only as a corporate body."[134] Or put another way, "it is the same evidence that is appealed to in justifying belief in the mathematical components of the theory that is appealed to in justifying the empirical portion of the theory."[135]

Maddy is critical of the indispensability thesis in that she sees a contradiction between the naturalist respect for the scientific method and holism.[136] In particular, she notes that working scientists have a wide range of attitudes towards well-confirmed scientific theories, which "vary from belief, through tolerance, to outright rejection".[137] This attitude is incompatible with the all-or-nothing approach which confirmational holism requires. Similarly she notes that

> Scientists seem willing to use strong mathematics whenever it is useful or convenient to do so, without regard to the addition of new *abstracta* to their ontologies, and indeed, even more surprisingly, without regard to the additional physical structure presupposed by that mathematics. On the

---

129 However, for a critique of the second proposition, see Hartry H. Field, *Science Without Numbers: A Defence of Nominalism* (Blackwell, 1980). Fields' program has itself been widely criticised. See for example Penelope Maddy, "Physicalistic Platonism" in A.D. Irvine (ed), *Physicalism in Mathematics* (Kluwer, 1990) at 259–289; David Malament, "Review of Field's Science Without Numbers" (1982) 79(9) *Journal of Philosophy* 523; Michael D. Resnik, "How Nominalist is Harry Field's Nominalism" (1985) 47(2) *Philosophical Studies* 163; Stewart Shapiro, "Conservativeness and Incompleteness" (1983) 80(9) *Journal of Philosophy* 521.

130 Colyvan, above n127.

131 Colyvan, above n127.

132 Colyvan, above n127.

133 Specifically confirmational holism, also called the Duhem-Quine thesis. See Colyvan, above n127.

134 W.V. Quine, "Two Dogmas of Empiricism" (1951) 60 *The Philosophical Review* 20 at 38.

135 Colyvan, above n127. See also Hilary Putnam, "What is Mathematical Truth", in *Mathematics Matter and Method: Philosophical Papers*, Volume 1 (2nd ed, Cambridge University Press, 1979) at 74: "[M]athematics and physics are integrated in such a way that it is not possible to be a realist with respect to physical theory and a nominalist with respect to mathematical theory."

136 Penelope Maddy, "Indispensability and Practice" (1992) 89(6) *Journal of Philosophy* 275; Penelope Maddy, "Naturalism and Ontology" (1995) 3(3) *Philosophia Mathematica* 248. See also Penelope Maddy, *Naturalism in Mathematics* (Clarendon Press, 1997).

137 Maddy, "Indispensability and Practice", above n136 at 280.

one hand, they do not subject these mathematical and structural hypotheses to testing; on the other, they do not regard the empirical success of a theory using strong mathematics as confirming the mathematical or structural hypotheses involved.[138]

Sober attacks the empirical basis of confirmational holism by arguing that the mathematics used in science is not subject to the same testing as the empirical aspects of scientific theories.[139] Further, the absence of alternatives to mathematics in supporting empirical theories suggests that mathematics is not confirmed by the empirical evidence. Despite these criticisms, however, "the debate is very much alive, with many recent articles devoted to the topic."[140]

Fictionalism: Conceived by Field[141] as a a response to Quine and Putnam's indispensability argument,[142] Field argued that whilst mathematics is undeniably useful, it is not essential to our understanding of the physical world.[143] To prove its dispensability, Field constructed a non-mathematical account of physics. Such an approach is interesting because it sidesteps the the issue of how mathematical propositions can be verified other than by empirical observation. Field's account, whilst recognised as a "major intellectual achievement",[144] has been widely criticised.[145]

Social constructivism: Social constructivists[146] see mathematics as merely "a social construction, a cultural product, fallible like any other branch of knowledge".[147] On this view, truth in mathematics depends on "mathematical

---

138 Maddy, "Naturalism and Ontology" above n136 at 255.
139 Elliot Sober, "Mathematics and Indispensability" (1993) 102(1) *Philosophical Review* 35.
140 Colyvan, above n127.
141 Field, above n129.
142 Field, above n129 at 5.
143 Field, above n129 at 7–8.
144 Shapiro, above n66 at 237.
145 For a summary of criticisms, see Mark Colyvan "Fictionalism in the philosophy of mathematics" in E.J. Craig (ed), *Routledge Encyclopedia of Philosophy Online edition*, (Taylor and Francis, 2011) <http://homepage.mac.com/mcolyvan/papers/fictionalism.pdf> (11 March 2011). For a different criticism, noting a problem analogous to that which Gödel's incompleteness theorem caused for the formalist programme, see Shapiro, above n66 at 235–236.
146 The leading accounts of social constructivism in mathematics are Hersh and Ernest. See Reuben Hersh, *What is Mathematics, Really?* (Oxford University Press, 1997); and Paul Ernest, *Social Constructivism as a Philosophy of Mathematics* (State University of New York Press, 1998). For a summary of Hersh and Ernest's views, and the difference between the two see Julian C. Cole, "Mathematical Domains: Social Constructs?" in Bonnie Gold and Roger Simons (eds), *Proof and Other Dilemmas: Mathematics and Philosophy* (Mathematics Association of America, 2008). The social constructivist account is influenced by the philosophies of Wittgenstein, who saw mathematics as a language game, and Popper, who saw mathematics as "an evolutionary product of the intellectual efforts of humans": Eduard Glas, "Mathematics as Objective Knowledge and as Human Practice" in Reuben Hersh (ed), *18 Unconventional Essays on the Nature of Mathematics* (Springer, 2006) at 289.
147 Paul Ernest, "Social Constructivism as a Philosophy of Mathematics: Radical Constructivism Rehabilitated?" (1990) <http://people.exeter.ac.uk/PErnest/soccon.htm> (10 September 2011), cited in Gold and Simons (eds), above n146 at 39.

traditions, methods, problems, meanings and values into which mathematicians are enculturated – that work to conserve the historically defined discipline".[148] In other words, mathematics is "constructed or created by – made real by – the activities of mathematicians".[149] On such an understanding, the source of mathematical truth is "neither physical nor mental, it's social. It's part of culture, it's part of history, it's like law, like religion, like money, like all those very real things which are real only as part of collective human consciousness."[150]

Social constructivism is a kind of neo-Kantianism in which the objectivity of mathematics is sourced neither in the platonic realm nor in the physical universe. Not surprisingly, this view is not warmly received by some mathematicians who object to the notion that mathematical knowledge is merely relative, rather than a truly objective discipline. Others note that the statement that mathematics is a human activity is trivial, and adds nothing to the debate.[151] Azzouni puts forward a more nuanced objection, suggesting that an account based on social practices alone fails to explain the unique degree of conformity in mathematics:

> I'm sympathetic to *many things* those who self-style themselves 'mavericks' have to say about how mathematics is a social practice. ... But many activities are similarly (epistemically) social: politicians ratify commonly-held beliefs and behaviour; so do religious cultists, bank tellers, empirical scientists and prisoners. ... It's widely observed that, unlike other cases of conformity, and where social factors *really are* the source of that conformity, one finds in mathematical practice *nothing like* the variability found in cuisine, clothing, or metaphysical doctrine.[152]

Structuralism: Structuralists[153] such as Shapiro and Resnik see mathematics as "the science of structures".[154] The view might be called quasi-platonist, in that structuralists assert that "each unambiguous sentence of [mathematics] is true or

148 Wikipedia, "Philosophy of Mathematics", <http://en.wikipedia.org/wiki/Philosophy_of_mathematics> (20 April 2007).

149 Cole, above n146 at 111.

150 John Brockman, "What Is Mathematics? A Talk With Reuben Hersh" *Edge* 5, 10 February 1997, <http://edge.org/documents/archive/edge5.html> (6 December 2010).

151 See for example Hacking, who levels an attack against social constructivist accounts of the philosophy of science as a failing to precisely define the claims which such a philosophy is making. Ian Hacking, *The Social Construction of What?* (Harvard University Press, 1999).

152 Jody Azzouni, "How and Why Mathematics is Unique as a Social Practice" in Reuben Hersh (ed) *18 Unconventional Essays on the Nature of Mathematics* (Springer, 2006) at 201–202.

153 See for example Michael D. Resnik, *Mathematics as a Science of Patterns* (1997, Clarendon Press); Shapiro, above n42. The emergence of structuralism is sometimes traced to Paul Benacerraf, "What Numbers Could Not Be" (1965) 74(1) *Philosophical Review* 47. A summary of the structuralist philosophy of mathematics can be found in Shapiro, above n66 ch 10; Erich H. Reck and Michael P. Price, "Structures and Structuralism in Contemporary Philosophy of Mathematics" (2000) 125 *Synthese* 341.

154 Shapiro, above n66 at 257.

false, independent of the language, mind, and social conventions of the mathematician."[155] But structuralists deny that mathematics is composed of independent objects such as numbers, arguing that "[t]he objects of mathematics, that is, the entities which our mathematical constants and quantifiers denote, are structureless points or positions in structures. As positions in structures, they have no identity or features outside of a structure."[156] The appeal of structuralism is that it corresponds with a greater emphasis on structures in modern mathematical practice.[157] But what is meant by the term structure? Various structuralist accounts are "significantly different from each other, even conflicting in many ways".[158] Further, each particular variant has its own philosophical problems, many of which are similar in nature to the issues facing the philosophies so far discussed.[159]

1.2   Can these various views be reconciled?

It should be apparent that all these accounts are impossible to fully reconcile with each other,[160] and such a task is not attempted here. The more modest goal sought, and an important task for grounding the mathematics exception, is to set out a view of the non-patentability of mathematics which accommodates all of these perspectives. In other words, a post-modern[161] holistic approach is taken.[162] This concedes that there is no one narrative which will provide the

155 Shapiro, above n66 at 257.
156 Michael D. Resnik, "Mathematics as a Science of Patterns: Ontology and Reference" (1981) 15 *Noûs* 529 at 530.
157 Reck and Price, above n153 at 346.
158 Reck and Price, above n153 at 374.
159 For example, an obvious question arises as to what the nature of these structures is. These structures are variously defended on platonist, formalist, and quasi-empiricist bases, introducing the same objections which those philosophies have sustained. For an overview of the main variants, and their individual difficulties, see Reck and Price, above n153.
160 For example, one cannot be both a platonist, believing that mathematical objects exist in a place accessible only by the mind, and an empiricist, believing that mathematical truth is verified only by correspondence with the physical world. Nor can one be a logicist, believing mathematics is a branch of logic, and an intuitionist, believing that mathematics must be experienced to be verified. Some combinations *are* possible, for example, noting Russell's combination of logicism and formalism. But a complete reconciliation remains impossible.
161 The author confesses to balking at describing the approach as post-modern, out of concern that post-modernism's distaste for meta-narratives, combined with the fact that it is itself a meta-narrative, is as paradoxical as a barber who shaves all those who do not shave themselves. Nonetheless, "postmodern is as good a name as any, especially since it's a bit of a joke on the ordinary meaning of modern. Obviously the Modern period was misnamed.": Larry Wall, "Perl, the first postmodern computer language" <http://www.perl.com/pub/1999/03/pm.html> (7 July 2011).
162 In the present context, it is submitted that an inclusive course it to be preferred over a skeptical one. To quote Wittgenstein, "[w]e just *can't* investigate everything, and for that reason we are forced to rest content with assumption": Ludwig Wittgenstein, *On Certainty* (G.E.M Anscombe and G.H. von Wright, (trans), Wiley-Blackwell, 1975). The approach

ultimate answer to the question of what mathematics is, but that each offers an insight into the nature of mathematics which captures something of the essence of mathematics, in the same way each of the four blind men captured something of the essence of the elephant.

Some support for a holistic view can be found in the attitudes of working mathematicians, since

> most mathematicians work in their respective fields 'doing mathematics' and concern themselves very little with questions of philosophy. Each one has formulated an opinion about what constitutes mathematics that is sufficient to guide him in his research, and these opinions are often mixtures.[163]

A good example of these mixtures can be seen in the philosophies of Gottlob Frege, who can be simultaneously identified with platonism,[164] logicism, transcendental idealism,[165] and whose work also greatly influenced formalism. Similarly, Heyting said:

> There is no conflict between intuitionism and formalism when each keeps to its own subject, intuitionism to mental constructions, formalism to the construction of a formal system, motivated by its internal beauty or by its utility for science and industry.[166]

Along similar lines, Avigad and Reck[167] suggest that despite Gödel's theorems, the formalist school still has a role to play in modern mathematics, in that it attempts to reconcile "general conceptual reasoning about abstractly characterized mathematical structures, on the one hand, and computationally explicit reasoning about symbolically represented objects, on the other".[168]

It is submitted that the philosophy of mathematics, is like mathematics itself,

> not a static body of revealed truth, but a complex of concepts in various stages of evolution, each of which is related to and affects all the others

---

might be thought of as a variation of Quine's web theory: "The totality of our so-called knowledge or beliefs ... is a man-made fabric which impinges on experience only along the edges.": Quine, above n134 at 39.

163 Berlinghoff, above n6 at 204.

164 "[T]he thought ... which we express in the Pythagorean theorem is timelessly true, true independently of whether anyone takes it to be true. It needs no bearer. It is not true for the first time when it is discovered, but is like a planet which, already before anyone has seen it, has been in interaction with other planets.": Gottlob Frege, "The Thought: A Logical Inquiry" (1956) 65 *Mind* 289 at 302.

165 Shabel, above n70 at 120, citing in support Gottlob Frege, *On the Foundations of Geometry and Formal Theories of Arithmetic* (Yale University Press, 1971).

166 Heyting, above n107 at 89.

167 Jeremy Avigad and Erich H. Reck, "Clarifying the nature of the infinite: the development of metamathematics and proof theory," (Carnegie Mellon Technical Report No CMU-PHIL-120, 11 December 2001).

168 Avigad and Reck, above n167, at 4.

and contributes to their growth. By a process of consolidation and generalization, these concepts frequently merge and are submerged in more all-embracing concepts that emphasize what the mathematician frequently refers to as "the unity of mathematics."[169]

The benefit of the post-modern holistic approach is that it provides a way forward out of the less familiar territory of mathematical philosophy, back to the more comfortable realm of law. On this approach, a theory of non-patentability is considered to be a success, not because it accords with one particular view of mathematics but because it can be substantially reconciled with *all* philosophies.

With this in mind, it is time to assess the way in which mathematics has been considered by patent law, to determine whether a theory of non-patentability exists which can be reconciled with all theories of mathematics.

2   A patent lawyer's account of mathematics

Although the European Patent Convention ("EPC") expressly excepts mathematical methods 'as such' from patentability in Article 52, no such categorical exception exists in either the US or Australia.[170] The approach adopted in those jurisdictions is to address the patentability of mathematics by reference to other notions of non-patentability. So although there may be agreement as to the inherent non-patentability of mathematics, it will be seen that patent law offers no cohesive explanation as to why that is so.

It may be that the difficulty stems from the rarity of finding a patent directed solely towards mathematical subject matter. More often, the patentability of mathematics arises in the context of computer-implemented methods, data manipulation, signal processing, and simulation/modelling used in industrial design processes. In any event, if the heart of what is claimed lies in a mathematical innovation or advance, then it is necessary to consider the effect of any claimed exclusion of mathematics from the scope of patentability.

Some of the territory covered in this section will be familiar, given that it overlaps with the case law on computer-related inventions canvassed in Chapter 2. However, the focus here is different, since the aim is to find an existing foundation on which to rest the non-patentability of mathematics which is consistent with the state of the law of patentable subject matter in the jurisdictions so far surveyed.

---

169 Raymond M. Wilder, "The Nature of Modern Mathematics," in William E. Lamon (ed), *Learning and the Nature of Mathematics* (Science Research Associates, 1972) at 47.
170 Even within the EU, at the EPO mathematical subject matter could still be claimed so long as it involved the use of a pen and paper, since this would satisfy the technicality requirement. See T258/03 *Hitachi/Auction Method* [2004] EPOR 55 ("*Hitachi*") at [4.6]. With the majority in *Bilski v Kappos* 130 S. Ct. 3218 (2010) ("*Bilski*") rejecting a categorical business method exception as 'atextual', the fate of a categorical mathematics exception would seem to be sealed in that jurisdiction as well.

2.1 Europe

Under Article 52 of the EPC, mathematical methods are not patentable *as such*. Article 52 forms part of the definition of patentable subject matter in s1(2)(c) *Patents Act 1977* (UK).

However, the words 'as such' in Article 52(2) have been the mechanism by which other exceptions, most notably the computer program exception, have been watered down by the European Patent Office ("EPO"). In particular, the "any hardware" approach reduces the Article 52 exclusions to mere form requirements when claimed as implemented on a computers,[171] since any method "when implemented by a computer or other hardware apparatus ... is a man-made technical apparatus having a utilitarian purpose, and not a method at all, and for that reason avoids exclusion under article 52(2)."[172]

A European Patent Office

The position of the EPO on the patentability of mathematics is as follows:

> These are a particular example of the principle that purely abstract or intellectual methods are not patentable. For example, a shortcut method of division would not be patentable but a calculating machine constructed to operate accordingly may well be patentable.[173]

The 2007 EPO case of *Circuit Simulation/Infineon Technologies*[174] illustrates how mathematical methods, when run on a computer, are patentable following the "any hardware" approach. The invention in that case was for a "computer-implemented method for the numerical simulation of a circuit".[175] It was claimed by the appellant inventor that the numerical simulation should be considered to meet the required standard of technical contribution on the basis that "technical considerations are required to solve problems in the engineering sciences, in particular electrical engineering ... to predict the performance of a circuit whose variables are technical parameters."[176] The appellant further claimed that "the simulation ... constitutes a technical process in itself"[177]

---

171 Or in any physical way. In *Hitachi* at [4.5] the EPO put it thus: "What matters having regard to the concept of "invention"... is the presence of technical character which may be implied by the physical features of an entity or the nature of an activity, or it may be conferred to a non-technical activity by the use of technical means."

172 Justine Pila, "Dispute over the Meaning of 'Invention' in Article 52(2) EPC – The Patentability of Computer-Implemented Inventions in Europe" (2005) 36 *International Review of Intellectual Property and Competition Law* 173 at 179.

173 European Patent Office, "Guidelines for Examination in the European Patent Office," December 2007, <http://www.epo.org/patents/law/legal-texts/guidelines.html> (13 January 2008) at 2.3.3.

174 T1227/05 *Circuit Simulation/Infineon Technologies* [2007] OJ EPO 574 ("*Infineon*").

175 *Infineon* at 575 (Claim 1).

176 *Infineon* at 577.

177 *Infineon* at 578.

since it created a model which "earlier technical literature holds to be difficult if not impossible".[178] The appellant also claimed a technical contribution on the basis that the instruction in the claims "is addressed to the technical engineer, not the mathematician, and thus in itself constitutes technical teaching".[179] Finally, the appellant claimed a technical contribution on the basis that this improved approach to modelling "requires shorter computing times and less storage space"[180] when implemented on a computer.[181]

The Board reasoned that "[a]s the method according to independent claim 1 or 2 is computer-implemented, it uses technical means and by that very token has technical character."[182] In arriving at this conclusion, the Board was "persuaded that [the] simulation … constitute[d] an adequately defined technical purpose for a computer-implemented method, provided that the method [was] functionally limited to that technical purpose".[183] The technical purpose was to be found in the claimed simulation of "a circuit with input channels, noise input channels and output channels whose performance is described by differential equations".[184] As the veracity of the simulation could be verified by "the physical and mathematical derivation specified in the system",[185] the board was "persuaded that the independent method claims [were] functionally limited".[186]

The Board also held that the invention was "neither a mathematical method as such nor a computer program as such, even if mathematical formulae and computer instructions are used to perform the simulation".[187] This was so because although the simulation could be performed by a human, as a practical matter, the "simulation method cannot be performed by purely mental or mathematical means".[188] Such simulations were "typical of modern engineering work"[189] and the increased efficiency of the claimed approach "enables a wide range of designs to be virtually tested and examined for suitability before the expensive circuit fabrication process starts."[190] As such,

> computer simulation methods for virtual trials are a practical and practice-oriented part of the electrical engineer's toolkit. What makes them so

178 *Infineon* at 578.
179 *Infineon* at 578.
180 *Infineon* at 578.
181 Cf. *Gale's Patent Application* [1991] RPC 305 ("*Gale's Application*"), discussed below.
182 *Infineon* at 581.
183 *Infineon* at 582.
184 *Infineon* at 582.
185 *Infineon* at 582.
186 *Infineon* at 582. See also 587, where the Board relevantly noted that the claims "both entail the specific modelling of an adequately defined class of technical systems (circuits) and define specific measures, not just mental constructs, for targeted implementation and application of the circuit model under the technically relevant conditions of 1/f noise."
187 *Infineon* at 582.
188 *Infineon* at 583.
189 *Infineon* at 583.
190 *Infineon* at 583.

important is that as a rule there is no purely mathematical, theoretical or mental method that would provide complete and/or fast prediction of circuit performance.[191]

On this basis, the claimed method was held to have the requisite technical character. The decision is interesting on a number of fronts. First, it illustrates the way that the any hardware approach makes almost any conceivable subject matter within the realm of the patentable. Second, the Board in its reasoning read into the "mathematical method" exclusion a requirement that the method be carried out by a human operator, thus excluding its relevance to computer-implemented methods. Finally, the patentability of mathematical innovations is clearly made dependent upon the audience to which such innovations are directed – if the innovation is likely to be of more interest to engineers than to mathematicians, then it is technical in character, and hence patentable.[192]

B   United Kingdom

As was seen in Chapter 2, the UK position is less amenable to the patenting of mathematical methods, although the position is not, technically at least, locked in.[193] According to the four-step approach of *Aerotel*,[194] where the technical contribution is to the field of mathematics, such an invention will fall foul of the Article 52 exclusion. In other words, there will be a narrow focus on the nature of the advance when determining the application of the patentable subject matter exclusions. Such an approach, which attempts to divine the substance of the invention is inconsistent with the approach adopted at the EPO, where any mention in the claims of a physical manifestation, or technical purpose, will render the claimed subject matter patentable. In *Symbian,* this inconsistency was acknowledged, with the Court seemingly torn between maintaining consistency with the EPO on the one hand, but also looking to avoid "throwing the law into disarray"[195] by adopting an interpretation which would mean that the exclusions have "lost all meaning".[196] There was some hope that some clear guidance might be garnered from the EPO by a reference of the issue to the Enlarged Board of Appeal.[197] However, the G3/08 referral avoided determining

---

191 *Infineon* at 583.
192 This reflects the claimed dichotomy between pure and applied mathematics, the difficulties of which was discussed above.
193 The Court of Appeal in *Symbian Ltd v Comptroller-General of Patents* [2008] EWCA Civ 1066; [2009] RPC 1 ("*Symbian*"), clearly indicated that the UK Courts would maintain their position until such time as "tolerably clear guidance" emerged from the EPO. Such guidance was no doubt expected from the G3/08 referral discussed below.
194 *Aerotel Ltd v Telco Holdings Ltd (and others) and Macrossan's Application* [2007] RPC 7 ("*Aerotel*").
195 *Symbian* at [46].
196 *Symbian* at [46].
197 G3/08 *Programs for computers* (Opinion of 12 May 2010) ("G3/08").

the issue, and the clash between the UK and EPO interpretations has remained unresolved in the years since.[198]

Gale's Application: *Gale's Application*, discussed in Chapter 2,[199] concerned an improved method of calculating the square root of a number, claimed as a ROM circuit designed to give effect to the method. On appeal to the Patents Court, Aldous J noted that

> the claim goes on to define [the ROM's circuitry] by the way it will be operated which in effect is a mathematical method of obtaining the square root of a number. No doubt the basis behind the claim can be said to be a mathematical method or a method for performing a mental act or even a program for a computer in that the ROM functions as a carrier or program which will be used in a computer."[200]

Despite this, his Honour held that because the claims were directed to a ROM, that this was different to a claim to the method itself, being

> a manufactured article having circuit connections which enables the program to be operated. A claim to a ROM with particular circuitry, albeit defined by functional steps, cannot to my mind be said to relate to the program or the functional steps as such.[201]

The Court of Appeal however rejected this interpretation, calling the claims to hardware a "confusing irrelevance",[202] noting that "in substance, a claim to a series of instructions which incorporate Mr. Gale's improved method of calculating square roots".[203] The Court went on to reject Mr Gale's invention on a number of related grounds:

> In the present case Mr Gale claims to have discovered an algorithm. Clearly that, as such, is not patentable. It is an intellectual discovery which, for good measure, falls squarely within one of the items, mathematical method, listed in section 1(2).[204]

The fact that it was a mathematical method was not dispositive of the issue. Nicholls LJ went on as follows:

> [T]he nature of the discovery is such that it has a practical application, in that it enables instruction to be written for conventional computers in a way

198 It might be that Brexit means this conflict ends with an amicable separation.
199 See Chapter 2 on page 60.
200 *Gale's Application* at 316 per Aldous J.
201 *Gale's Application* at 317 per Aldous J.
202 *Gale's Application* at 326 per Nicholls LJ.
203 *Gale's Application* at 326 per Nicholls LJ.
204 *Gale's Application* at 327 per Nicholls LJ.

which will, so it is claimed, expedite one of the calculations frequently made with the aid of a computer. In my view the application of Mr. Gale's mathematical formulae for the purpose of writing computer instruction is sufficient to dispose of the contention that he is claiming a mathematical method as such.[205]

However, the Court concluded that the invention could not escape the operation of the computer program exclusion, since the method neither embodied "a technical process which exists outside the computer",[206] nor did the instructions "solve a 'technical' problem lying within the computer".[207] Although the Court did not explain this conclusion further, the conclusion that the improved algorithm, although it "makes a more efficient use of a computer's resources"[208] did not solve a "technical" problem is interesting. It is submitted that the lack of technicality refers the inquiry back again to the mathematics exception – an intellectual discovery or mathematical method such as that claimed cannot be technical.

Citibank v Comptroller of Patents: The patentability of mathematical methods was directly considered in the *Citibank* case.[209] The claims under consideration concerned a method of managing risks associated with trade in financial derivatives by ensuring the integrity and validity of data used in such an evaluation.[210] The claimed method used a statistical analysis to compare the current data set with historical data to determine "the likelihood that changes to the set of input data are the result of one or more errors."[211]

Counsel for the Comptroller asserted that the claims were for a method, the substance of which was "two calculations and a statistical comparison and analysis".[212] Counsel for Citibank "sought to draw a distinction between mathematical methods on the one hand and methods of calculation involving the application of mathematical methods on the other",[213] with the difference between the two said to reside in the fact that the former existed at a similarly "high level" to discoveries and scientific theories.[214]

Mann J noted that no authority as to the meaning of the mathematical method exclusion was placed before the Court. After reviewing *Gale* and *Fujitsu*, his Honour noted that those cases did not support the sort of distinction sought to be made by Citibank, and held that the claims were a mathematical method

---

205 *Gale's Application* at 327 per Nicholls LJ.
206 *Gale's Application* at 327 per Nicholls LJ.
207 *Gale's Application* at 328 per Nicholls LJ.
208 *Gale's Application* at 327 per Nicholls LJ.
209 1 (09 June 2006) ("*Citibank*").
210 *Citibank* at [2].
211 *Citibank* at [3].
212 *Citibank* at [19].
213 *Citibank* at [20].
214 *Citibank* at [20]. What was meant by this "high level" was not further explained, but it seems safe to suggest that it is a reference to the abstract nature of the subject matter collected in this exclusion.

"both in terms of the normal use of language and in terms of the likely policy underlying the Patents Act".[215]

Mann J also flagged the relationship between mathematical methods and methods of performing mental acts, remarking that "the interrelationship between the concepts of a mental act and mathematical calculation might have assisted in arriving at a resolution of this matter, but am left with the appeal as presented to me."[216]

2.2   The US

In contrast to the EU approach, it will be recalled that the codification of US patent law in §101 contains no express exceptions, instead using a positive definition of four types of patentable subject matter, namely processes, machines, manufactures, and compositions of matter.

A   Early jurisprudence

US jurisprudence on the patentability of mathematics demonstrates the slippery nature of mathematics. As determined above, there is considerable disagreement between mathematicians about what mathematics is. This uncertainty is also manifest in the case law. Some judgements attempt to equate mathematics with algorithms;[217] in other instances mathematics is seen as "imperfect proxies for mathematical truths and other laws of nature."[218] For example, the 1939 US Supreme Court case of *Mackay Radio*[219] focused on the communicative role mathematics often plays in science, and set in place the conceptual division between 'discoveries' and their 'applications'. In this case, the US Supreme Court held that

> [w]hile a scientific truth, or the mathematical expression of it, is not patentable invention, a novel and useful structure created with the aid of knowledge of scientific truth may be.[220]

The applicant in that case sought to alter the mathematical formula describing his invention to in order to cover a new configuration of antenna said to infringe his

---

215 *Citibank* at [26].
216 *Citibank* at [29]. In obiter his Honour dismissed a claim that for practical purposes the method would need to be run on a computer as not relevant. This was because the claims were clearly directed to mental acts, since they referred "to methods and techniques, not the physical means": at [29].
217 Most notably see *Gottschalk v Benson* 409 US 63 (1973), discussed in Chapter 2. See the criticism therein by Newell, noting that it encourages an illusory distinction between the numerical and non-numerical.
218 John A. Burtis, "Towards a Rational Jurisprudence of Computer-Related Patentability in Light of *In re Alappat*," (1995) 79 *Minnesota Law Review* 1129 at 1157.
219 *Mackay Radio v RCA* 306 US 86 (1939) ("*Mackay Radio*").
220 *Mackay Radio* at 94.

patent. The court refused the application on the grounds that where a scientific principle is expressed as a mathematical formula, altering the formula is not allowed where doing so would alter the nature of the law on which it was based.[221] This court in this case clearly captured the notion of mathematics as a language to describe the attributes of nature. As such, it could be traced back to a bar on the patenting of scientific principles, as discussed in *O'Reilly v Morse*.[222]

B Yuan

In the case of *Yuan*,[223] the issue of patentability and mathematics arose in a different context. In that case, the appellant had applied for a patent on a high-speed airfoil with low drag characteristics.[224] However, the court determined that the appellant's contribution lay "in a mathematical procedure by which the aircraft designer can start with a pressure distribution curve of the required characteristics and convert it into a velocity distribution curve".[225] As such, the invention was said to comprise "purely mental steps dependent upon the mathematical formula which is recited in, and constitutes the heart of, the claims".[226] Since it had been "thoroughly established by decisions of various courts that purely mental steps do not form a process which falls within the scope of patentability as defined by statute",[227] the subject matter of the invention was not patentable.[228]

C Benson and Diehr

The patentability of mathematics received its most detailed treatment in the US jurisdiction in the computer software cases. In the landmark case of *Gottschalk v Benson*,[229] it was "held that the discovery of a novel and useful mathematical formula may not be patented".[230] The claimed invention was "a faster and more efficient mathematical procedure for transforming the normal 'decimal' type of numbers (base 10) into true 'binary' numbers (base 2) which are simpler to process within computers.[231] The inventors claims were "not limited to any particular art or technology, to any particular apparatus or machinery, or to any particular end use",[232] although in argument before the court, the inventors' attorney stated that the claim did not extend to its use by a human using a pen and paper.

221 *Mackay Radio* at 98.
222 *O'Reilly v Morse* 56 US 62 (1854).
223 *In re Yuan* 188 F.2d 377 (1951) ("*Yuan*").
224 *Yuan* at 378.
225 *Yuan* at 379.
226 *Yuan* at 380.
227 *Yuan* at 380.
228 On the patentability of mental steps, see also *In re Abrams* 89 USPQ (BNA) 266 (1951). The mental steps doctrine was covered in more detail in Chapter 2, Section 2 on page 40.
229 409 US 63 (1973).
230 *Parker v Flook* 437 US 584 (1978) ("*Flook*") at 585.
231 Oyez, "Gottschalk v Benson" <http://www.oyez.org/cases/1970-1979/1972/1972_71_485/> (26 Feb 2008).
232 *Gottschalk v Benson* at 64.

In deciding the instant invention was non-patentable, the Court invoked earlier case law on the non-patentability of scientific principles,[233] and also the non-patentability of ideas.[234] The Court concluded that "the 'process' claim [was] so abstract and sweeping as to cover both known and unknown uses"[235] and that as a result, the practical effect of allowing the patent would be that "the patent would wholly preempt the mathematical formula and, in practical effect, would be a patent of the algorithm itself."[236]

This position was supported in the later Supreme Court case of *Diamond v Diehr* where the court said that "an algorithm, or mathematical formula, is like a law of nature, which cannot be the subject of a patent."[237]

D   Alappat and State Street

Despite this seeming support for mathematics as non-patentable, inferior courts, and in particular the Federal Circuit, advanced the patentability of both software and mathematics by exploiting theoretical weaknesses in the *Benson* and *Diehr* judgements. As the Federal Circuit noted in *Alappat*:

> The Supreme Court has not been clear ... as to whether such subject matter is excluded from the scope of 101 because it represents laws of nature, natural phenomena, or abstract ideas. See *Diehr*, 450 U.S. at 186 (viewed mathematical algorithm as a law of nature); *Gottschalk v. Benson*, 409 U.S. 63, 71–72 (1972) (treated mathematical algorithm as an "idea"). The Supreme Court also has not been clear as to exactly what kind of mathematical subject matter may not be patented. The Supreme Court has used, among others, the terms "mathematical algorithm," "mathematical formula," and "mathematical equation" to describe types of mathematical subject matter not entitled to patent protection standing alone. The Supreme Court has not set forth, however, any consistent or clear explanation of what it intended by such terms or how these terms are related, if at all.[238]

In *Alappat*, the court made it clear that it did not consider that mathematics constituted another category of non-patentable subject matter:

> Mathematics is not a monster to be struck down or out of the patent system, but simply another resource whereby technological advance is achieved.[239]

---

233 See *Gottschalk v Benson* at 67–68.
234 "An idea, of itself, is not patentable.": *Rubber-Tip Pencil Company v Howard* 87 US 498 (1874), cited in *Gottschalk v Benson* at 67–68.
235 *Gottschalk v Benson* at 68.
236 *Gottschalk v Benson* at 72.
237 *Diehr* at 186.
238 *In re Alappat* 33 F.3d 1526 (1994) ("*Alappat*") at 1543, footnote 19.
239 *Alappat* at 1570.

As a result of this confusion, the law of patentable subject matter has moved on to a point where many more recent decisions are hard to reconcile with a mathematics exception, such as *State Street*,[240] where transformation of numerical data, put through a series of mathematical calculations by a computer, constitutes patentable subject matter so long as it provides a "a useful, concrete, and tangible result"[241] without pre-empting other uses of the algorithm. Much of this encroachment into the patentability of mathematics was achieved through the inclusion of a computer as a tangible component of the invention which executed the algorithm. However, in *AT&T v Excel* the need to claim a computer was removed; any application of mathematics which produced a "useful, concrete and tangible result" would be sufficient.[242] That is, the inclusion of a computer in the claims, to run the algorithm, was no longer necessary. In the wake of *Bilski* however, the continued applicability of this approach must be doubted.

The court clearly noted that

> nothing in today's opinion should be read as endorsing interpretations of §101 that the Court of Appeals for the Federal Circuit has used in the past. See, e.g., *State Street*, 149 F. 3d, at 1373; *AT&T Corp.*, 172 F. 3d, at 1357. It may be that the Court of Appeals thought it needed to make the machine-or-transformation test exclusive precisely because its case law had not adequately identified less extreme means of restricting business method patents, including (but not limited to) application of our opinions in *Benson*, *Flook*, and *Diehr*.[243]

E Pre-Bilski practice

Until 2010, the Eighth Edition of the USPTO Manual of Patent Examining Procedure[244] ("MPEP") maintained a separate section for the patentability of mathematics. That has since been removed, with the patentability of mathematics now being addressed by the *Mayo* two-step approach. The Eighth Edition did not

---

240 *State Street Bank & Trust Co v Signature Financial Group Inc* 149 F.3d 1368 (1998) (*"State Street"*).

241 *State Street* at 1373.

242 "Whether stated implicitly or explicitly, we consider the scope of §101 to be the same regardless of the form – machine or process – in which a particular claim is drafted.": *AT&T Corp v Excel Communications Inc* 172 F.3d 1352 (1999) (*"AT&T"*) at 1357.

243 *Bilski* at 3223 per Kennedy J. Stevens J went further, noting that relying on the useful effect approach would be a "grave mistake": at 3232. Similarly, Breyer J noted that the rejection of the machine-or-transformation test as the sole test for patentability "by no means indicates that anything which produces a 'useful, concrete, and tangible result,' is patentable.": at 3259 per Breyer J (citations omitted).

244 United States Patent and Trademark Office, "2106.02 Mathematical Algorithms," Manual of Patent Examining Procedure, Eighth Edition, August 2001, Revision 5, August 2006 <http://bitlaw.com/source/mpep/2106_02.html> (21 August 2017).

adopt the useful result approach and suggested instead that claimed inventions involving mathematics will be non-statutory (non-patentable) where they:

- consist solely of mathematical operations without some claimed practical application (i.e., executing a "mathematical algorithm"); or
- simply manipulate abstract ideas, e.g., a bid (*Schrader*, 22 F.3d at 293–94, 30 USPQ2d at 1458–59) or a bubble hierarchy (*Warmerdam*, 33 F.3d at 1360, 31 USPQ2d at 1759), without some claimed practical application.[245]

Following the Eighth Edition, it would seem that the only limit to the patentability of a mathematical innovation is one's ability to imagine a practical application for it. Mathematicians have long noted the impossibility of drawing distinctions between pure and applied mathematics, in that almost every area of mathematics, no matter how "pure" it may have been considered at one time or another, has been found to have a practical application. The MPEP also goes on to explain the relationship between laws of nature and mathematical algorithms in the following terms:

> Certain mathematical algorithms have been held to be nonstatutory because they represent a mathematical definition of a law of nature or a natural phenomenon. For example, a mathematical algorithm representing the formula $E = mc^2$ is a "law of nature" – it defines a "fundamental scientific truth" (i.e., the relationship between energy and mass). To comprehend how the law of nature relates to any object, one invariably has to perform certain steps (e.g., multiplying a number representing the mass of an object by the square of a number representing the speed of light). In such a case, a claimed process which consists solely of the steps that one must follow to solve the mathematical representation of $E = mc^2$ is indistinguishable from the law of nature and would "pre-empt" the law of nature. A patent cannot be granted on such a process.[246]

Klemens noted that despite continuing theoretical support for the exclusion of mathematics "standing alone", the US judiciary had, prior to *Bilski* at least, enabled the patenting of mathematics. He gives examples of the following four patents:

- Method and system for solving linear systems (U.S. Patent No. 6078938).
- Cosine algorithm for relatively small angles (No. 6434582).
- Method of efficient gradient computation (No. 5886908).
- Methods and systems for computing singular value decompositions of matrices and low rank approximations of matrices (No. 6807536).[247]

245 MPEP "2106.02 Mathematical Algorithms".
246 MPEP "2106.02 Mathematical Algorithms".
247 Ben Klemens, "Software Patents Don't Compute," *IEEE Spectrum Online*, July 2005, <http://spectrum.ieee.org/careers/careerstemplate.jsp?ArticleId=i070305> (4 January 2008).

F Bilski

It will be recalled from Chapter 2 that the invention claimed in *Bilski* was "a method of hedging risk in the field of commodities trading".[248] *Bilski* must therefore be taken to reinforce the non-patentability of mathematics, as the subject matter of the claims was clearly a mathematical algorithm, including data gathering and processing, without any claims to a physical instantiation designed to carry the method out. The invention was characterised by the majority as being directed to a non-patentable abstract idea.[249] Further, the existence of a "useful result" will no longer be of relevance, as the *State Street* approach was clearly rejected.[250]

Previous USPTO practice has changed significantly since *Bilski*.[251] The post-Bilski Manual of Patent Examining Procedure guidelines specifically recite mathematical concepts as "nothing more than abstract ideas".[252]

The Manual does leave open the possibility of patenting machine-implemented mathematical algorithms, although this is likely to depend on the extent to which the "elements of the claim, considered both individually and as an ordered combination, are sufficient to ensure that the claim as a whole amounts to significantly more than [mathematics] itself."[253] Similarly, the transformation limb of the "machine-or-transformation" test does allow the possibility of patenting computer-implemented mathematics, to the extent that the implementation would transform a general purpose computer into a specific-purpose computer, along the lines discussed in Chapter 2.[254]

2.3 Australia

Mathematics is not included in the express exceptions in *Patents Act 1990* (Cth) s18, nor was it included in the *Statute of Monopolies 1623* (UK) on which the Australian definition of patentability relies.[255] Nevertheless, "mathematical algorithms ... have traditionally been regarded as not per se patentable, because they do not exhibit the requirements of a manner of manufacture."[256]

---

248 *In re Bernard L. Bilski and Rand A. Warsaw* (2008) 545 F.3d 943 ("*In re Bilski*") at 949.
249 *Bilski* at 3230 per Kennedy J.
250 *Bilski* at 3231 per Kennedy J.
251 United States Patent and Trademark Office, "Interim Guidelines for Determining Subject Matter Eligibility Claims in View of *Bilski v Kappos*" (2010) 75(143) *Federal Register* 43,922.
252 United States Patent and Trademark Office, *Manual of Patent Examining Procedure*, "2106 – Patent Subject Matter Eligibility" <https://www.uspto.gov/web/offices/pac/mpep/s2106.html> (21 August 2017).
253 MPEP, "2106 – Patent Subject Matter Eligibility".
254 United States Patent and Trademark Office, "Interim Guidelines for Determining Subject Matter Eligibility Claims in View of *Bilski v Kappos*" (2010) 75(143) *Federal Register* 43,922 a 43,925, specifically factor B(3).
255 Section 18 (1)(a) requires that for an invention to be patentable it must be "a manner of manufacture within the meaning of section 6 of the Statute of Monopolies".
256 Australian Patent Office, "2.9.2.5 Discoveries, Ideas, Scientific Theories, Schemes and Plans" in *Manual of Practice and Procedure* <http://www.ipaustralia.gov.au/pdfs/patentsmanual/WebHelp/Patent_Examiners_Manual.htm> (10 September 2011).

Australian patentable subject matter law has been consistently moving away from express categorisations towards a general test for patentability.[257] The most likely determinant of the patentability of a mathematical innovation depends on whether it is, as a matter of substance rather than form,

- an artificially created state of affairs, and
- of utility in the field of economic endeavour.[258]

Where what is sought to be patented is a process, "it must offer some advantage which is material, in the sense that the process belongs to a useful art."[259] The focus of the inquiry in the context of mathematics is often a question of whether the mathematical algorithm in question produces a 'useful effect'. The most recent statement of this requirement comes from *Grant v Commissioner of Patents*,[260] where the Full Court of the Federal Court held:

> It has long been accepted that … a mathematical algorithm … *without effect* [is] not patentable. … It is necessary that there be some "useful product", some physical phenomenon or effect resulting from the working of a method for it to be properly the subject of letters patent.[261]

Neither *RPL Central* nor *Research Affiliates* dealt squarely with the patentability of mathematics. However, the adoption in those cases of UK and US jurisprudence as consistent with the Australian approach, and the reasoning apposite,[262] it is a small step to see how this "useful product" or "physical effect" approach might be equated with a "technical effect" or technicality requirement.[263]

The rejection of the patentability of a legal structure, and the divination of a requirement of physicality which accords with the *Bilski* position noted above seem to suggest that the non-patentability of mathematics is likely to continue in this jurisdiction. However, this is not a certainty, as the position in Australia

---

257 In *Research Affiliates LLC v Commissioner of Patents* [2014] FCAFC 150 ("*Research Affiliates*") [116]: "The approach to be taken to deciding whether a claimed method or product is properly the subject of letters patent must be flexible and must allow for new technologies presently unknown. The principles should be applied irrespective of the area of human endeavour and invention under consideration." See also William van Caenegem, *Intellectual Property* (2nd ed, LexisNexis Butterworths, 2006) at 126–127.

258 Cf. *Research Affiliates* at [101]: "The characterisation of patentability by reference to only the description in [*National Research Development Corporation v Commissioner of Patents* (1959) 102 CLR 252 ("*NRDC*")] of a product which consists of an artificially created state of affairs of economic significance was part of the High Court's reasoning but did not represent a sufficient or exhaustive statement of the circumstances in which a claimed invention is patentable."

259 *Research Affiliates* at [101].

260 (2006) 154 FCR 62 ("*Grant*").

261 *Grant* at 73 (emphasis added).

262 *Research Affiliates* at [59].

263 *Commissioner of Patents v RPL Central* [2015] FCAFC 177 ("*RPL Central*") at [99].

mirrors that of the US as espoused in *State Street*. If nothing else, *IBM v Commissioner of Patents*,[264] held by the Full Federal Court in *Grant* to be consistent with their physical result requirement, is in fact a patent on purely mathematical subject matter. In that claimed invention, a set of numerical control points were put through a couple of known mathematical algorithms to produce a table of numbers which could be put into memory to produce a curve on a screen. This was held to be a manner of manufacture as it produced a commercially useful effect in the field of computer graphics.

The inputs to the claimed algorithm were as follows:

> A set of control points which define the curve and which are input for each dimension and a number of intervals of the curve to be computed.[265]

The aspect of the algorithm most focused on however was the output, or effect:

> [I]t is not suggested there is anything new about the mathematics of the invention. What is new is the application of the selected mathematical methods to computers, and in particular, to the production of the desired curve by computer. This is said to involve steps which are foreign to the normal use of computers and, for that reason, to be inventive. The production of an improved curve image is a commercially useful effect in computer graphics.[266]

The problem with accepting this analysis is that the actual display of the curve on the screen of a computer is taken care of by an entirely independent set of components, namely the video card (display adapter). To understand why this is the case, a basic understanding of how things are drawn on the screen is necessary. Each of the dots on a computer screen, or pixels, is represented in the memory of the computer as a single value. The screen is represented in a contiguous series of memory addresses which map out a 'virtual screen' table corresponding to the dots on the table. A simple 9×9 monochrome (single colour) screen can be represented by a 10×10 table of pixel values, with 0 being black and 1 being white. A blank screen would be represented as shown in Table 3.1. By updating the values stored in the table, a simple circle could be represented by changing some of the pixel values in memory as shown in Table 3.2 (the zeros are represented by blank cells).

At the time of the case, a typical screen size was 1024×768 pixels, with a range of 65536 colours available for each pixel (called 16-bit colour depth). To accommodate this, a much larger table is required, but the concepts remain the same. Colours are handled by storing an integer between 0 and 65535 in the table. A

264 (1991) 33 FCR 218 ("*IBM (AU)*").
265 *IBM (AU)* at 4 (Claim 1).
266 *IBM (AU)* at 225–226.

Table 3.1 Blank screen

0	0	0	0	0	0	0	0	0
0	0	0	0	0	0	0	0	0
0	0	0	0	0	0	0	0	0
0	0	0	0	0	0	0	0	0
0	0	0	0	0	0	0	0	0
0	0	0	0	0	0	0	0	0
0	0	0	0	0	0	0	0	0
0	0	0	0	0	0	0	0	0
0	0	0	0	0	0	0	0	0

Table 3.2 A circle on the screen

				1	1	1	1			
			1						1	
		1								1
		1								1
		1								1
		1								1
			1						1	
				1	1	1	1			

Table 3.3 16-bit colour representation

15	14	13	12	11	10	9	8	7	6	5	4	3	2	1	0

16-bit pixel value actually contains the binary values of three colour channels – red, green and blue, as shown in Table 3.3.

After the display memory is set, the display adapter reads the virtual screen data. At the time of IBM's case, these digital pixel values would have to be then passed through a digital-to-analogue converter built in to the video card:

The modern display comes from a long linage of cathode ray tubes (CRTs). A CRT display uses an electron gun to blast three different materials on the inside of the tube that emit red, green and blue light when excited. These early devices were analog by nature and to convert from digital to analog a device called a digital to analog converter (DAC) made its way into graphics outputs.[267]

A 'virtual screen' table such as that produced by the IBM algorithm could through similar means be sent to a printer instead of a screen. The operation of the display adapter and digital analog converter are entirely independent of the claimed algorithm here, and it is wrong to treat them as a part of the claimed invention. If this element is not considered to be part of the patented invention, then we are left with a process which has as its inputs, a series of numbers representing control points of a curve. These inputs are put through a number of mathematical transformations to arrive at a table of integer numbers. Clearly such an invention is in essence mathematical in nature.

2.4   Why is mathematics non-patentable?

Armed with the requisite knowledge of the nature of mathematics, derived from the history and philosophies of mathematics discussed above, it might be hoped that it would be a simple matter to find a suitable (or at least consistent) basis upon which to rest the non-patentability of mathematics. There are numerous bases upon which it might be sought to set such a claim. The case law considering mathematics frames the discussion in various ways, including:

* discoveries, or intellectual discoveries;[268]
* scientific theories, or the expression thereof;[269]
* an intellectual discovery;[270]
* abstract information, abstract ideas;[271] and
* mental steps.[272]

---

267 Don Woligroski, "The Basic Parts of a Graphics Card" *Tom's Hardware*, 24 July 2006 <http://www.tomshardware.com/reviews/graphics-beginners,1288–2.html> (15 October 2010).

268 The notion of mathematics as an non-patentable discovery extends back at least as far as to 1884, when in *Young v Rosenthal* (1884) 1 RPC 29 at 31, Grove J referred to mathematics as an "abstract discovery". Mathematics has been couched in similar terms since that time. This correspondence between discoveries and mathematics is enshrined in Article 52 of the EPC, where mathematical methods are listed together with discoveries and scientific theories. See also *Citibank* at [20]; *Gale's Application* at 327 per Nicholls LJ.

269 *Mackay Radio* at 94.

270 *Gale's Application* at 327 per Nicholls LJ.

271 See *Bilski* at 3230 per Kennedy J. Cf. *Alice Corp v CLS Bank International* 134 S.Ct. 2347 (2014) ("*Alice Corp*") at 2357, where the Court noted that the fact a the method in *Bilski* could be reduced to a mathematical formula did not assume "talismanic significance". See also *Citibank*.

272 *Yuan*.

It might be thought that one (or various) of these descriptions, arising as they do from the exposition of the concept of patentable subject matter over time in the jurisdictions considered, would be possible candidates upon which to base the non-patentability of mathematics. However, most of these explanations are problematic for one reason or another.

Some of the claims meet immediate objections based on the nature of mathematics discussed in Section 1. For example, the classification of mathematics truths as concerning scientific truths or discoveries, whilst consistent with "realist" philosophies such as empiricism and platonism, cannot be reconciled with constructivist accounts, which if accepted would suggest that mathematics is created, and hence may be patentable where of economic use. The latter position raises an issue of whether mathematics can properly be considered to be something "under the sun that is made by man"[273] or an "artificially created state of affairs".[274] This suggests that the notion of discovery might be synonymous with the notion of abstract idea or an absence of technical character.

Similarly, attempting to distinguish between discoveries, theories, intellectual information or abstract ideas and their application is problematic because of the interrelationship of so-called 'pure' and 'applied' mathematics:

> What is considered applied mathematics today may, by a curious reversal of process, become pure mathematics tomorrow. And at any given moment of time, there is no clear distinction between what is pure and what is applied. I have even noticed how two groups of mathematicians, each of which considers itself applied, have each denied the propriety of the designation "applied" to the the fellow group. And ... even the mathematician who insists he is a pure mathematician is in reality an applied mathematician in that his interests are applications to the conceptual world of mathematics.[275]

Similarly, Lobachevsky said that "[t]here is no branch of mathematics, however abstract, which may not some day be applied to phenomena of the real world."[276] The non-Euclidean geometries such as that which Lobachevsky developed are a case in point.[277] Similarly, as noted by Atiyah above, mathematics at the

---

273 *Diamond v Chakrabarty* 447 US 303 (1980) at 309.
274 *NRDC* at 277.
275 Raymond M. Wilder, "The Nature of Modern Mathematics," in William E. Lamon (ed), *Learning and the Nature of Mathematics* (Science Research Associates, 1972) at 46.
276 Cited in George E. Martin, *The foundations of geometry and the non-Euclidean plane* (Springer, 1998) at 225.
277 "No more impressive warning can be given to those who would confine knowledge and research to what is apparently useful, than the reflection that conic sections were studied for eighteen hundred years merely as an abstract science, without regard to any utility other than to satisfy the craving for knowledge on the part of mathematicians, and that then at the end of this long period of abstract study, they were found to be the necessary key with which to attain the knowledge of the most important laws of nature.": Alfred North Whitehead, *Introduction to Mathematics* (Williams & Northgate, 1911) at 136–137.

turn of the 21st century is greatly informed by, and therefore closely related to, modern developments in physics.[278]

Even the most seemingly impressive accounts have problems. The interrelationship between mathematical methods and mental steps was referred to favourably by the Court in the UK case of *Citibank*,[279] and was dispositive of the issue in the US case of *Yuan*. This doctrine showed great promise in the software context, in that it brought into issue the relationship between the human mind and the computer. Given that mathematics is largely performed without a computer, and noting the prominence of thought processes inherent in the notions of abstraction, deduction, and proof discussed above, it might be thought that such a doctrine would be particularly useful in addressing the non-patentability of mathematics. But the abandonment of the doctrine in the US in favour of a focus on "mathematical algorithms" means that any argument against patentability built on such a basis stands on shaky ground.[280]

The now dominant notion in the current US approach, that of the non-patentable abstract idea,[281] might appear to be the mental steps doctrine in a different guise. Even if it is not, this approach, which seems to acknowledge the fundamental role which abstraction plays in mathematics, is not without problems. The notion of abstraction, and the corresponding machine-or-transformation 'clue' call into question the importance of manifesting invention in a physical form. As Stevens J noted in *Bilski*, where the claims were dismissed as non-patentable abstract ideas, the majority failed to provide "a satisfying account of what constitutes an non-patentable abstract idea".[282] Nor can that answer be found in *Alice*. This author would add to Stevens J's criticism, that there exists a failure to adequately explain *why* such abstract ideas should remain non-patentable. Perhaps this is the reason why a distinction between discovery and invention "is not precise enough to be other than misleading".[283] The lack of precision in these concepts arises from a lack of proper explanation of the policy which informs them.

In all the jurisdictions just surveyed, it seems to be accepted that, in theory if not in practice, mathematics is not patentable. And in its favour, the mathematics

---

278 Claims to a clear distinction between pure and applied science in general have been questioned by Stokes, who notes an additional class of use-inspired basic research. See Donald E. Stokes, *Pasteur's Quadrant – Basic Science and Technological Innovation* (Brookings Institution Press, 1997).

279 Mann J noted at [27] that "closeness of those concepts was reflected by Laddie J in *Fujitsu Limited's Application* [1996] RPC 511, 532" and that to his Honour's mind, "the interrelationship between the concepts of a mental act and mathematical calculation might have assisted in arriving at a resolution of this matter": at [30].

280 Although the original formulation of the doctrine is quote old now, more recent post-*Bilski* cases have given it a new lease of life: see for example *CyberSource Corporation v Retail Decisions Inc* Appeal No 2009-1358 (Fed Cir, 2011). The limitations of the mental steps approach are discussed further in Chapter 6.

281 *Alice Corp.*

282 *Bilski* at 3236.

283 *NRDC* at 264.

exception has lasted nearly 400 years. But the practical reality is at odds with the theory. Given the long-term trend towards expansionism, and the demise of the business method exception both in Australia and in the US, the lack of a well understood mathematics exception is a cause for concern, especially given the usefulness of mathematical advances to modern product markets such as:

- bioinformatics
- computing
- cryptography
- finance
- robotics
- image processing
- nanotechnology.

This concern is particularly acute when one considers the fate of the business method exception. Although not expressly specified, such an exception was held to exist for a long time.[284] However, in the US and Australia, this category of patentable subject matter was first undermined, and then removed. In *State Street*, as noted in Chapter 3, the business method exception was described as "ill-conceived"[285] and summarily dismantled. Although *State Street* has been similarly wiped out by *Bilski v Kappos* and later Supreme Court decisions, the majority's textual approach leaves no room for a resurrection of this categorical exception.[286] There is similarly nothing in the text of the the US *Patents Act* or Constitution which bars the patenting of mathematics.

In Australia, Heerey J in *Catuity*,[287] following the US approach from *State Street*, held that there was no reason to maintain an exception for business methods, preferring instead that applications for patents in this area be judged upon the standard criteria for patentability in line with *NRDC*. If similar logic were to be applied to mathematics, then a mathematical advancement could be considered patentable so long as it is described by reference to a "concrete, tangible, physical or observable effect".[288] Given how IBM was demonstrated to be directed to mathematical subject matter above, this suggests that the alleged physical effect in that case amounted to a mere form requirement,

---

284 "Although it is difficult to derive a precise understanding of what sorts of methods were patentable under English law, there is no basis in the text of the *Statute of Monopolies*, nor in the pre-1790 English precedent, to infer that business methods could qualify": *Bilski* at 3240 per Stevens J. In relation to the subsequent development of US law, Stevens J, after a detailed analysis noted that "the historical clues converge on one conclusion: A business method is not a 'process'": at 3250.

285 *State Street* at 1375.

286 "The Court is unaware of any argument that the 'ordinary, contemporary, common meaning,' *Diehr* ... at 182, of 'method' excludes business methods.": *Bilski* at 3221 per Kennedy J.

287 *Welcome Real-Time SA v Catuity Inc and Ors* (2001) 113 FCR 110.

288 *Grant* at 70.

despite valiant attempts to bring it within the *Research Affiliates* and *RPL Central* dedication to substance over form.[289]

This mismatch between the theory and reality arises from a fundamental misunderstanding of the reason why mathematics stands outside the patent regime. If the reasons for excluding mathematics are not well understood, then it is to be expected that support for it will, at worst, be gradually eroded. At best, our understanding of the patentability of mathematics might fluctuate like the ebb and flow of the tide. But the very persistence of the exception to until more recent times suggests that it reflects important factors at work in the patent regime. Through attempting to highlight the reasons why mathematics is considered non-patentable, this book will attempt to highlight some of these inner workings of the patent system and further inform debate about the dimensions of patent law. Further chapters will attempt to reconcile the differences between a mathematical and patent law understanding of mathematics, by looking not to the question of what mathematics *is* but to what mathematics *requires* – the optimal conditions for its advancement.

3 Conclusion

This chapter has compared a mathematical understanding of the nature of mathematics, derived from the historical and philosophical accounts of mathematics, with a legal understanding of the nature of mathematics as set out in case law in the three jurisdictions considered throughout this thesis.

Whilst it has been made clear that the issue of the nature of mathematics is not one which is likely to be resolved, it is through understanding the various accounts together that the limitations of present explanations of mathematics' non-patentability can be seen. It has been demonstrated that no explanation thus far put forward is acceptably wide that it can accomodate all present understandings of mathematics.

As such, the next chapter builds upon this understanding of the nature of mathematics and the limitations of previous accounts of its non-patentability, by looking at the issue in a different fashion – not at what mathematics is, but at what it requires for mathematical innovation to prosper. This is surely the right question, since it is towards the promotion of that innovation to which patent law is said to be directed. The next chapter will use these historical and philosophical accounts of the nature of mathematics just considered to build a set of common features found therein. From these commonalities, it will be argued why mathematical innovation depends on freedom to continue to advance. From that understanding, an explanation of its non-patentability will be advanced.

---

289 The Full Federal Court in *RPL Central* were right to observe that the "method in [*IBM (AU)*] could have been characterised simply to involve 'drawing a curve on a computer'": at [105]. It is the author's contention that this is exactly what the patent in that case should have been characterised as.

4 Why mathematics is not patentable

There was a blithe certainty that came from first comprehending the full Einstein field equations, arabesques of Greek letters clinging to the page, a gossamer web. They seemed insubstantial when you first say them, a string of squiggles. Yet to follow the delicate tensors as they contracted, as the superscripts paired with subscripts, collapsing mathematically into concrete classical entities – potential; mass; forces vectoring in a curved geometry – that was a sublime experience. The iron fist of the real: inside the velvet glove of airy mathematics.[1]

1 Introduction

It has been seen that legal accounts of the nature of mathematics considered thus far do not sit well with the various philosophies of mathematics. Some explanations fall short of the mark, and some, whilst showing promise and capturing some aspect of the nature of mathematics, give at best a partial account. This chapter proposes to take a different approach. Whereas the historical, philosophical, and legal accounts of mathematics considered have centred on what mathematics *is*, it will be seen that the better focus, is on what mathematics *requires* for its further progress. It will be shown that by looking at the requirements of mathematicians, that is, by focusing on the nature of mathematical development, or mathematical innovation, it is possible to avoid difficult questions about the ontological status of mathematical objects. In that way it is possible to avoid the impossible task of providing a definitive unifying account of what mathematics is, but nonetheless to answer the question of whether mathematics is a proper subject for patent protection.

The basis of mathematical progress is freedom, namely the ability of mathematicians to create new mathematics free of constraints. Without that freedom, the advancement of mathematics would be impossible. More specifically, the mathematics exception should be understood as a recognition of the utmost importance of freedom of thought and expression to the continued development of

---

1 Gregory Benford (1941) in Lloyd Albert Johnson, *A Toolbox for Humanity: More than 9000 Years of Thought* (Trafford Publishing, 2004) at 89.

mathematics. It will be shown that this freedom is evident from the cognitive and expressive aspects which lie at its core.

It is not suggested, however, that the need for freedom distinguishes mathematics from other fields. To an extent, freedom is a prerequisite of progress in all fields of human endeavour. However, it is the *extent* of the freedom required which makes mathematics distinctive from other fields falling within the ambit of patentable subject matter. This chapter therefore explores the role which freedom plays in mathematics, leading back towards a proper explanation of its non-patentability, consistent with traditional notions of patentability. What is asserted is that the level of freedom required by mathematics in order for it to advance is greater than the level of freedom available within the patent paradigm.

In exploring the role of freedom in mathematical advancement, three specific aspects of that freedom are noted. First, the importance of freedom of thought underscores the abstract, rather than physical, nature of mathematics. Second, the need for freedom of expression is borne out by the notion of mathematics as a language. As a language, mathematics is expressive rather than purposive. Although mathematics may be put to work in useful processes, its essence lies its ability to express our understanding of those processes. In other words, it is the symbolic nature of mathematics that lies at its core. Third, the need for expression also draws out the aesthetic rather than rational nature of mathematics. The end product of mathematical activity, mathematical proofs, present as sequences of logical deductions from accepted axioms. But this logical, rational nature belies the creative process by which it is forged.

It is these three aspects, the abstract, expressive, and aesthetic, which pave the road back to patent law. In simple terms, it will be shown that mathematics is not to be patentable because it has the characteristics of the fine arts, rather than the useful arts. It will also be shown that this explanation of non-patentability can be reconciled with philosophies of mathematics discussed in Chapter 3. This explanation of mathematics' non-patentability, and the analytical tools therein derived, will be applied to an analysis of the patentability of software in the next chapter.

2   Freedom in patent law

The contemporary understanding of the role of patent law is as an incentive to innovation.[2] The mechanism by which patent law seeks to incentivise innovation is through the promise of monopoly rights in exchange for disclosure of technological advances.[3] The monopoly right awarded to the patent grantee,

2 "The essence of the patent system is to encourage entrepreneurs to develop and commercialise new technology": Commonwealth of Australia, "Patents Bill 1990: Second Reading" Senate, 29 May 1990.
3 See *Liardet v Johnson* (1778) 1 Carp Pat Cas 35 (NP) ("*Liardet*"); *R.C.A. Photophone Ltd. v Gaumont-British Picture Corporation* (1936) 53 RPC 167 ("*RCA Photophone*") at 1 per Romer LJ.

being in the nature of a proprietary right to exclude, is based upon a paradigm of control. A consequence of this grant of control, is a loss of freedom of competitors, who are excluded from the area of invention defined by the boundaries of the patent. Thus patent law can be characterised as an attempt to balance "conflicting public interests of a proper reward for investors on the one hand without an unreasonable fetter on the freedom of third parties on the other".[4] To say this is to draw attention to the social contract theory of patent law, put forward in *Liardet*, which suggests loss of those freedoms is the cost borne by the public in exchange for the award of monopolies to encourage advancement. The emergence of a social contract theory of patent law at the same time as broader notions of a social contract,[5] and the importance of the historical investigation of manner of manufacture, suggests that reference to freedom as understood now and then is appropriate. However, a structural bias exists in patent law, which has a tendency to focus attention on the former interest at the expense of the latter:

> We know that [law] built around the self-interest of existing and aspirant monopolists will protect a variety of private goods, namely those of the firms and interests at the table. We know also that it will fail to protect certain kinds of interests – most notably those of large numbers of unorganised individuals with substantial collective, but low individual, stakes in the matters being discussed.[6]

Using the paradigm of freedom of developed in human rights jurisprudence to define and draw attention to the interests of these unrepresented many, is therefore an important way of addressing the imbalance. It should however be noted that the reference to human rights jurisprudence will not be used to expound a reconception of intellectual property based entirely in human rights, as as a paradoxical competition between a right to property versus other human rights.[7] Rather, an understanding of the impact of monopoly on those other than the inventor/creator must be understood as one effect of the award of patent rights. What is relied upon is the explanatory power of these notions to aid

4   Thomas Terrell and Simon Thornley, *Terrell on Patents* (15th ed, Sweet & Maxwell, 2000) at 13.
5   Social contract theory was first put forward by Thomas Hobbes, in *Leviathan* (1651). According to Hobbes, it is through an act of consent to a social contract wherein each individual "lay down this right to all things; and be contented with so much liberty against other men, as he would allow other men against himselfe" and thereby created a civil society. See also Locke, *Second Treatise of Government* (1689), VIII, §99, for a varied account.
6   James Boyle, "Enclosing the Genome: What the Squabbles over Genetic Patents Could Teach Us" in F. Scott Kieff, *Perspectives on the Human Genome Project* (Academic Press, 2003) 97 at 117. See also Peter Drahos with John Braithwaite, *Information Feudalism: Who Owns the Knowledge Economy?* (Earthscan Publications, 2002).
7   As to why, see Rochelle C Dreyfuss, "Patents and Human Rights: Where is the Paradox" in William Grosheide (ed), *Intellectual Property and Human Rights: A Paradox* (Edward Elgar, 2010) 72 at 73.

in the provision of a satisfactory account of a class of excluded subject matter in the context of the patentable subject matter inquiry.

3 What is meant by freedom?

Freedom is a concept which has attracted the attention of countless poets, philosophers, and statesmen. But the concern of this thesis is not this broad conception of freedom as "the quality or state of being free",[8] but freedom as "a political or civil right".[9] In particular, it is the jurisprudence of human rights, namely freedom of thought, and the closely related right of freedom of expression which require further investigation.

3.1 Freedom of thought

A useful starting point is the definition given to it in the *Universal Declaration of Human Rights*,[10] Article 18:

> Everyone has the right to freedom of thought, conscience and religion; this right includes freedom to change his religion or belief, and freedom, either alone or in community with others and in public or private, to manifest his religion or belief in teaching, practice, worship and observance.[11]

Philosophical justifications of freedom of thought find their source in the 18th century, during the Enlightenment, which was "characterised by a rejection of knowledge derived through tradition and authority, including religious authority. Instead, the fundamental source of knowledge became reason – that is, the capacity of human beings to know truth through independent and critical thought".[12] J.S. Mill was a strong adherent of freedom of thought, which he thought should include "absolute freedom of opinion and sentiment on all subjects, practical or speculative, scientific, moral or theological".[13] In many ways, freedom of thought is the font from which all other freedoms flow:

---

 8 "Freedom" in *Merriam-Webster's Dictionary of Law* (Merriam-Webster, 1996).
 9 "Freedom" in *Merriam-Webster's Dictionary of Law* (Merriam-Webster, 1996).
10 *Universal Declaration of Human Rights*, GA Res 217A (III), UN GAOR, 3rd sess, 183rd plen mtg, UN Doc A/810 (10 December 1948) ("the UDHR").
11 The right is similarly defined in Article 19, *International Covenant on Civil and Political Rights*, opened for signature 16 December 1966, 999 UNTS 171 (entered into force 23 March 1976).
12 Lawrence McNamara, "Chapter 1: Free Speech," in Des Butler and Sharon Rodrick, *Australian Media Law* (2nd ed, Lawbook Co, 2004) at 5. On rationalism in mathematics see Chapter 3 at 88, and below.
13 John Stuart Mill, *On Liberty* (J.W. Parker and Son, 1859) at 11.

> Freedom of thought ... is the matrix, the indispensable condition, of nearly
> every other form of freedom. With rare aberrations a pervasive recognition of
> this truth can be traced in our history, political and legal.[14]

As a consequence, it is unsurprising that a close relationship between freedom of
thought and freedom of expression exists.[15] In the introduction to *A History of
Freedom of Thought*,[16] Bury describes the relationship as follows:

> It is a common saying that thought is free. A man can never be hindered
> from thinking whatever he chooses so long as he conceals what he thinks.
> The working of his mind is limited only by the bounds of his experience
> and the power of his imagination. But this natural liberty of private thinking
> is of little value. It is unsatisfactory and even painful to the thinker himself, if
> he is not permitted to communicate his thoughts to others, and it is obvi-
> ously of no value to his neighbours. Moreover it is extremely difficult to
> hide thoughts that have any power over the mind. If a man's thinking
> leads him to call in question ideas and customs which regulate the behaviour
> of those about him, to reject beliefs which they hold, to see better ways of
> life than those they follow, it is almost impossible for him, if he is convinced
> of the truth of his own reasoning, not to betray by silence, chance words, or
> general attitude that he is different from them and does not share their opin-
> ions. Some have preferred, like Socrates, some would prefer today, to face
> death rather than conceal their thoughts. Thus freedom of thought, in
> any valuable sense, includes freedom of speech.[17]

The interconnectedness of thought and language, and indeed the proposition
that language determines thought,[18] was similarly expounded by J.S. Mill[19] and
popularised by George Orwell in his novel *1984*.[20] The link between thought
and speech was also explored by the psychologist Vygotsky, by studying the
development of children. He concluded as follows:

---

14 *Palko v State of Connecticut* 302 US 319 (1937) at 327.
15 Similarly, note also the inclusion of a freedom to *manifest* thought in the UDHR definition
    above.
16 John B. Bury, *A History of Freedom of Thought* (Williams and Norgate, 1914).
17 Bury, above n16 at 7–8.
18 Such a notion is known as linguistic determinism. See for example Benjamin Whorf,
    "Science and Linguistics" (1940) 42 *Technology Review* 229. See also Wittgenstein: "The
    limits of my language mean the limits of my world.": Ludwig Wittgenstein, *Tractatus
    Logico-Philosophicus* (Cosimo, 2010), at 88 (Proposition 5.6).
19 Mill, above n13 at 11–12: "The liberty of expressing and publishing opinions may seem to
    fall under a different principle, since it belongs to that part of the conduct which concerns
    other people, but, being almost of as much importance as the liberty of thought itself and
    resting in great part on the same reasons, is practically inseparable from it."
20 See George Orwell, "Appendix: The Principles of Newspeak", *Nineteen Eighty Four*,
    <http://www.netcharles.com/orwell/books/1984-Appendix.htm> (16 April 2008).

It would be wrong ... to regard thought and speech as two unrelated processes, either parallel or crossing at certain points and mechanically influencing each other. The absence of a primary bond does not mean that a connection between them can be formed only in a mechanical way ... The meaning of a word represents such a close amalgam of thought and language that it is hard to tell whether it is a phenomenon of speech or a phenomenon of thought. A word without meaning is an empty sound; meaning, therefore, is a criterion of "word," its indispensable component. It would seem then that it may be regarded as a phenomenon of speech. But from the point of view of psychology, the meaning of every word is a generalization or a concept. And since generalizations and concepts are undeniably acts of thought, we may regard meaning as a phenomenon of thinking.[21]

Because of the difficulties in actually directly restricting the thoughts of another, and also because of the lack of consensus on exactly what thought is,[22] it is difficult to make any authoritative pronouncements as to the nature of this freedom. Given the relationship between thought and expression just discussed, attention is often focused on the indirect protection of thought through the protection of expression.[23] However, as will be seen below, both aspects draw attention to different parts of the practice of mathematics, and as such will be separately considered.

3.2   Freedom of expression

Freedom of expression, sometimes referred to as freedom of speech, is also defined in the *UDHR*, in Article 19:

> Everyone has the right to freedom of opinion and expression; this right includes freedom to hold opinions without interference and to seek, receive and impart information and ideas through any media and regardless of frontiers.[24]

Freedom of expression can be justified by its relationship to a set of "basic purposes, aims, or goals thought to be pursued by our constitutional protection of

---

21 Lev S. Vygotsky and Alex Kozulin *Thought and Language* (2nd ed, MIT Press, 1986) at 211–212.

22 Debates as to what thought is are "as old as human beings and as fascinating as life itself." Anil K Rajvanshi, *Nature of Human Thought* (NARI, 2004) at 4.

23 This is also the approach of the European Court of Human Rights, who "prefer to examine an applicant's complaints solely under Article 10 [freedom of speech] if possible.": John Wadham et al., *Blackstone's Guide to the Human Rights Act 1998* (4th ed, 2007). See *Paturel v France*, App No 54968/00, 22 December 2005.

24 The link noted earlier between freedom of expression and freedom of thought is immediately apparent, by virtue of the inclusion of freedom of opinion. Article 19, *International Covenant on Civil and Political Rights*, opened for signature 16 December 1966, 999 UNTS 171 (entered into force 23 March 1976) is drawn in similar terms.

freedom of speech itself".[25] Whilst there may be many ways of categorising the various values which free speech is said to protect, the primary justifications are based on autonomy, truth, and democracy.[26] Each of these will now be considered in turn.

A   Truth

J.S. Mill's *On Liberty*[27] is the leading exposition of the notion of freedom of expression. Within a utilitarian framework,[28] he argued for an almost unbridled freedom of speech. This was based on "a sort of conceptual Darwinism: the conviction that in a 'free market of ideas', the best will come to the fore and survive".[29] Specifically, Mill offered three arguments for the importance of the free expression of ideas:

> First, he suggested that the idea suppressed as false may in fact be true, since to contend otherwise is to assume the infallibility of the individuals who adhere to the dominant opinion. Second, he argued that the suppressed opinion might be at least partially true, since one view rarely contains all of the truth in a given area. Finally, Mill suggested that even if the suppressed idea were completely false, its suppression would tend to result in the true idea's becoming a sterile and unchallenged dogma that would lack the vital force necessary for a living truth.[30]

This 'truth' goes beyond a scientific correctness, to include "political or ethical truths".[31] The value of the truth justification lies in the expression itself, wherein any "epistemic advance" has value.[32] As such, free speech puts value on "truths, half-truths, gross errors and vividly and emptily held truths of

---

25 R. George Wright, "Why Free Speech Cases Are As Hard (And As Easy) As They Are" (2001) 68 *Tenn. L. Rev.* 335 at 337.

26 See Lawrence McNamara, "Chapter 1: Free Speech," in Des Butler and Sharon Rodrick, *Australian Media Law* (2nd ed, Lawbook Co, 2004); Wright, above n25; Keith Werhan, *Freedom of Speech: A Reference Guide to the United States Constitution* (Greenwood Publishing Group 2004) at 28. As to Emerson's four justifications, see C. Edwin Baker, *Human Liberty and Freedom of Speech* (Oxford University Press, 1989) at 47; cf. Susan Hoffman Williams *Truth, Autonomy, and Speech: Feminist Theory and the First Amendment* (NYU Press, 2004) at 27–31.

27 John Stuart Mill, *On Liberty* (J.W. Parker and Son, 1859).

28 Mill argued that the best utilitarian outcome resulted from a minimal regulation of individual liberty: see Mill, above n27 at 9–10.

29 Richard Chappell, "On Liberty," *Philosophy, et cetera*, 29 June 2004, <http://www.philosophyetc.net/2004/06/on-liberty.html> (27 June 2008). See also McNamara, above n26 at 7.

30 Christopher Wonnell, "Truth and the Marketplace of Ideas" (1986) 19 *University of California at Davis Law Review* 669 at 671.

31 McNamara, above n26 at 7.

32 McNamara, above n26 at 8.

many sorts; in politics, culture and entertainment, as well as science; and in preparatory as well as public or final forms of expression".[33]

Critics of the truth theory have pointed out that it is by no means certain that truth will prevail over falsity,[34] that this assumption rests on a "false confidence in the rationality of human beings"[35] and the claim that there is such thing as objective truth.[36] Yet despite this, the truth theory continues to be an influential factor in free speech jurisprudence.

B Autonomy

Freedom of expression is also said to enhance the autonomy of the individual both as an influence, and as a capacity.[37] As an influence, it acts by enabling access to information which influences individual decision making, leading to a better use of their autonomy.[38] Similarly, having a wide variety of information available encourages self-reflection, since it allows an individual "to question and claim assumptions that they might otherwise simply accept uncritically and thereby increase their autonomy."[39] On a similar note, free expression may be seen to promote intellectual, moral and social capacities of individuals.[40] As a capacity, expression should be understood as "something that is directly and intrinsically valuable of itself, rather than being something that is instrumentally valuable".[41] On this account, expression is the means by which individuals give effect to their autonomy.[42]

For Kant, "autonomy is not only clearly recognisable as a free speech value for the sake of legally protected speech, but is also at the same time nothing less than 'the ultimate justification of the State.'"[43] This justification of free speech was also found in the judicial development of free speech by US

---

33 Wright, above n25 at 339.

34 See for example Wonnell, above n30 at 671–672; Frederick Schauer, *Free Speech: A Philosophical Enquiry* (1982) at 23; Stanley Ingber, "The Marketplace of Ideas: A Legitimizing Myth" [1984] *Duke Law Journal* 1.

35 Wonnell, above n30 at 672–673.

36 For a detailed discussion see Kent Greenawalt, "Free Speech Justifications" (1989) 89(1) *Columbia Law Review* 119 at 131–141; Wonnell, above n30 at 673–674.

37 Filimon Peonidis, "Freedom of Expression, Autonomy, and Defamation" (1998) 17(1) *Law and Philosophy* 1 at 3.

38 Williams, above n26 at 18.

39 Williams, above n26 at 18.

40 Williams, above n26 at 19.

41 McNamara, above n26 at 6.

42 Williams, above n26 at 19.

43 Wright, above n25 at 341. This is because freedom forms a necessary condition for Kant's categorical imperative, "[a]ct only according to that maxim whereby you can at the same time will that it should become a universal law": Immanuel Kant, *Grounding for the Metaphysics of Morals*, at 421, which Kant describes as "the canon for morally estimating any of our actions": Kant, at 424.

Supreme Court Justice Walter Brandeis. For example, in *Whitney v California*, his Honour said:

> Those who won our independence believed that the final end of the state was to make men free to develop their faculties … They believe that freedom to think as you will and to speak as you think are means indispensable to the discovery and spread of political truth … But they knew order cannot be secured merely through fear of punishment for its infraction; that it is hazardous to discourage thought, hope and imagination; that fear breeds repression; that repression breeds hate; that hate menaces stable government; that the path of safety lies in the opportunity to discuss freely supposed grievances and proposed remedies; and that the fitting remedy for evil counsels is good ones.[44]

C   Democracy

The final justification for free speech is based on its being a necessary condition for democracy. This is supported by a view that speech is information, and the more information an electorate is able to access, the better the quality of democratic representation they will choose.[45] However, the requirements of a functioning democratic system need not be limited to so narrow a range of expression. Democratic governance depends on the existence of a "a robust, pluralist civil society"[46] in which citizens develop "independent spirit, self-direction, social responsibility, discursive skill, political awareness, and mutual recognition"[47] in order for democratic culture to take hold. These preconditions themselves can be translated into free speech issues. For example, any true democratic system is built around the notion of one person, one vote. Translated into the realm of expression, this requires that "there are no significant inequalities of power when people communicate."[48] Thus "it could well be argued that the state should act to remove inequalities in the communicative process and ensure that the silenced are heard and the powerful do not prevail in public discourse."[49] This is consistent with Mill's caveat that "[l]iberty, as a principle, has no application to any state of things anterior to the time when mankind have become capable of being improved by free and equal discussion."[50]

Similarly, free speech helps build a culture of participation. This participation should not be limited to participation in the electoral process, but to the creation of a democratic culture, or a "culture in which people can participate actively in

44 *Whitney v California* (1927) 274 U.S. 357 at 375.
45 McNamara, above n26 at 9.
46 Neil Weinstock Netanel, "Copyright and a Democratic Civil Society" [1996] 106 *Yale Law Journal* 283 at 342.
47 Netanel, above n46 at 343.
48 McNamara, above n26 at 10.
49 McNamara, above n26 at 10. Also note the democratic requirement of equality above can also suggest that speech should be *limited* in certain situations, namely where "a democratic community agrees on a moral principle that should guide actions": *ibid*.
50 Mill, above n13 at 24.

the creation of cultural meanings that in turn constitute them".[51] Put another way, by allowing people to participate in the creation of culture, freedom of speech imbues individuals with an expectation that they should be allowed to participate in the selection of government, thereby underscoring the notion of popular sovereignty.[52]

A right of free expression based on this broader understanding of democracy extends well beyond the governmental sphere into the private sector:

> [The state] must facilitate the democratization of associational and communicative frameworks to provide greater opportunities for citizen engagement and self-government. Concomitantly, it must work to modify or eliminate social arrangements that undermine democratic citizenship while still leaving considerable room for "bottom-up" community organizing, education, and direction.[53]

Thankfully, it is not necessary to proffer a view on which of these characterisations of freedom of expression is the correct one. For present purposes it is sufficient to say that they help to form a backdrop against which an understanding of the impact of a patent grant can have on others. The truth justification aligns to the notion that individual contributions to an overlapping field of exploration are to be valued for the reason that any one of them might give rise to a valuable advance. As such, any restriction on freedom ought to consider the possibility that the award of control risks the possibility that a *better* invention might be lost, and the public interest ill served as a result. The autonomy theory makes clear that individual contributions are valuable in their own right, suggesting that the ability of individuals to act without restriction is something worth protecting, regardless of any inventive merit their entry into a particular sphere may bring about. Finally, the democratic theory of patent law underscores the fact that competition is important because it builds a culture of participation, and is as such inherently valuable in a democratic society.

4 Is freedom required in mathematics?

Having defined what is meant by freedom of thought and expression, the next task is to demonstrate how innovation in mathematics depends on mathematicians having such freedoms in order to advance their art. In the previous chapter, a

---

51 Jack M. Balkin, "How Rights Change: Freedom of Speech in the Digital Era" (2004) 26 *Sydney Law Review* 5 at 12.
52 Netanel, above n46 at 343. See also Marci A. Hamilton, "Power, Responsibility, and Republican Democracy" (1995) 93 *Michigan Law Review* 1539 at 1539–1540; Frank I. Michelman, "Law's Republic" (1988) 97 *Yale Law Journal* 1493 at 1500–1503; John Stuart Mill, "Chapter 2: The Criterion of a Good Form of Government" in *Considerations on Representative Government* (1861) <http://ebooks.adelaide.edu.au/m/mill/john_stuart/m645r/chapter2.html> (4 March 2011).
53 Netanel, n46 at 345.

number of philosophies of mathematics were introduced. In developing the role of freedom in mathematics it will be necessary to rely on the understanding developed therein. As such, a brief review of the theories may be useful.

Platonist theory posits that the realm of mathematics is made of abstract objects which exist independently of human minds, although it is through ill-defined mental faculty that they are accessed. The truth of a theory is therefore determined by comparing the theory with an "observation" of those objects. Despite being one of the oldest theories of mathematics, it is still relied on by working mathematicians, who treat the objects of mathematics as if they exist. The rationalist school holds that the objects of mathematics are purely mental creations, and the correctness of mathematical theories is ultimately reliant on human reason. The empiricist school in contrast posited that the correctness of mathematics ultimately depended on its validation through empirical observation of the real world. Kant's transcendental idealism is a mixture of rationalism and empiricism, which posits that it is through the combination of observation and intuition that truth could be tested.

Following the foundational crisis of mathematics towards the end of the 19th century, these older philosophies developed into a three-way fight for supremacy. The logicist school, a refined version of rationalism, held that all mathematics could be subsumed within logic, and attempts were made to demonstrate how all of arithmetic could be derived using rules of logical inference from a series of accepted axioms and definitions. A related school was the formalist school, which argued that mathematics was a game played using axioms and rules like those posited by the logicists. However, the choice of logic is, on a strictly formalist account, only an arbitrary one. The intuitionist school, a variation of Kant's transcendental idealism, posited that all mathematics was based on intuition derived from experience, although intuition plays the primary role.

In the 20th century, quasi-empiricism modified the empiricist position, accepting the abstract nature of mathematics, but tying its ultimate validity to empirical observation. Fictionalism asserted a new version of formalism, suggesting that the whole of mathematics was a convenient fiction. Social constructivists agreed it was a fiction, but suggested it could be verified by agreement between mathematicians. Structuralists sought to shift the debate, by replacing mathematical objects with patterns and structures, although the dependency on platonist, empiricist, rationalist, and formalist methods of verification remain.

With this potted history of mathematics in mind, a consideration of the role of thought and expression in mathematics is possible. The role of thought and expression, although acknowledged as interrelated, are again considered separately.

4.1   Thought in mathematics

The word mathematics comes from the Greek word μάθυμα (máthema) meaning "something learned"[54] and comes from "the same Indo-European base ... as

---

54 "Mathematics" in John Ayto, *Word Origins: the hidden histories of English words from A to Z* (A & C Black, 2006).

produced English memory and mind".[55] This definition alone is suggestive of the central role of thought in mathematics. Support for such claims can be found by looking at the descriptive terminology[56] applied to the mathematical subject matter such as "proof",[57] "deduction",[58] "logic",[59] "analysis",[60] "conjecture",[61] "hypothesis",[62] "theory",[63] "lemma",[64] and mathematical activities like "reasoning", "solving", "abstraction",[65] "generalisation", and "demonstration". As would be expected then, there are explicit descriptions of mathematics as being concerned with thought. Shapiro for example is in no doubt:

> The practice of mathematics is primarily a *mental* activity. To be sure, mathematicians use paper, pencils, and computers, but at least in theory these are

---

55 "Mathematics" in Ayto, above n54.

56 This approach was adopted in *Halliburton Oil Well Cementing Co v Walker* (1944) 146 F.2d 817 at 821, a case involving the mental steps doctrine (discussed in Chapter 3).

57 The dominant modern conception of proof in mathematics is as "a purely logical construct validated in purely syntactic terms": Leo Corry, "II.6 The Development of the Idea of Proof" in Timothy Gowers et al. (eds), *The Princeton Companion to Mathematics* (2008) at 140. However, Corry notes that such a notion of proof has limitations: at 141. In a different context, Atiyah relevantly notes that "[p]roof is the end product of a long interaction between creative imagination and critical reasoning": Sir Michael Atiyah, "VIII.6 Advice to a Young Mathematician" in Gowers et al.

58 "When the modern subject is studied as an abstract deductive science in its own right, it is often referred to more fully as pure mathematics": "mathematics", *OED Online,* November 2010 <http://www.oed.com/view/Entry/114974> (11 September 2011).

59 "The study of numbers, shapes, and other entities by logical means.": "mathematics" in R. David Nelson (ed), *The Penguin Dictionary of Mathematics* (2008); "a science (or group of related sciences) dealing with the logic of quantity and shape and arrangement": "mathematics", *Princeton WordNet* <http://wordnetweb.princeton.edu/perl/webwn?s=mathematics> (1 March 2011). Logic forms the core of the logicist theory of mathematical philosophy; as well as being a branch of mathematics. The development of logic as a branch of mathematics was described by George Boole in the Introduction to his book *An Investigation of the Laws of Thought* (Project Gutenberg, 2005) at 1. The role of logic, and the impact of Boole's project on computer science are discussed in Chapter 5.

60 Used as both a classification of the subject matter of mathematics, but also as part of the research methodology of mathematics in any area.

61 A conjecture is "[a] statement which may be true, but for which a proof (or disproof) has not been found": "conjecture (hypothesis)" in Nelson, above n59. On the role of conjectures in the development of mathematics, see Corry, above n57 at 142.

62 The term is used interchangeably with conjecture, an example being Riemann's hypothesis, which does not bear simple explanation. For more see "Riemann zeta function," in Nelson, above n59; Andrew Granville, "IV.2 Analytic Number Theory" in Gowers et al. (eds), above n57 at 337.

63 For a specific example, see "Cantor's theory of sets" in Nelson, above n59. The term is often used to describe branches of "pure" mathematics, such as number theory, knot theory, game theory, decision theory and information theory.

64 A lemma is "a subsidiary proposition that is assumed to be true in order to prove another proposition": "lemma", *Princeton WordNet* <http://wordnetweb.princeton.edu/perl/webwn?s=lemma> (1 March 2011).

65 In relation to the central role which abstraction plays in mathematics, see Chapter 3, Section 1 on page 77.

dispensable. The mathematician's main tool is her mind. Although the philosophies [of mathematics] are quite different from (and even incompatible with each other, they all place emphasis on this activity of mathematics, paying attention to its basis or justification.[66]

Further, many of the philosophies of mathematics clearly place mental processes at the core. This is nowhere more stark than in intuitionism, which describes mathematics as the process of "deducing theorems exclusively by means of introspective construction",[67] and thereby "a study of certain functions of the human mind".[68] The role of thought in the rationalist school is also clear, since adherents were "impressed with the seemingly unshakeable foundation enjoyed by all mathematics, and its basis in pure rationality. They tried to put all knowledge on the same footing."[69] Similarly, Frege's logicist programme[70] and Kant's transcendental idealism[71] also place obvious emphasis on the role of mental processes in mathematics.

   Other accounts of mathematics might be less obvious in their endorsement of the primary role of the mind in mathematics, but at the very least do not contradict such a claim. Recall from Chapter 3 the formalist characterisation of mathematics as a "meaningless game".[72] Accepting this claim in the strictest sense does not mean mental activity is not central to the formalist programme. The formulation of the rules of the system (the axioms and allowable deductions), the "playing" of the game, and the underlying goal of replacing the ambiguities of intuition in mathematics with a rigorous and coherent formal foundation all point towards the primacy of mental processes in the formalist account. If one accepts a less radical formalism, wherein it is admitted that the symbols have meanings, and the game results in logical consequences,[73] the significance of the mind is further amplified.

66 Stewart Shapiro, *Thinking About Mathematics* (Oxford University Press, 2000) at 172–3. Similarly, the definition put forward in Chapter 3 suggested mathematics is "a process of thinking that involves building and applying abstract, logically connected networks of ideas." <http://www.project2061.org/publications/bsl/online/ch9/ch9.htm#Mathema ticalWorld> (accessed 15 May 2007). See Chapter 3, at 78. See also Dirk J Struik, *A Concise History of Mathematics* (Dover Publications, 1967) at 1; Leonard Peikoff, *Objectivism: The Philosophy of Ayn Rand,* Penguin Group, New York 1991, at 90; "Mathematics" in Kenneth McLeish (ed), *Bloomsbury Guide to Human Thought* (Bloomsbury Publishing, 1993).
67 L.E.J. Brouwer, "Consciousness, Philosophy and Mathematics" (1948) cited in Shapiro, above n66 at 8.
68 Arund Heyting in *Intuitionism: An Introduction* (1956), cited in Shapiro, above n66 at 8.
69 Shapiro, above n66 at 3.
70 See Chapter 3, at 93.
71 See Chapter 3, at 90.
72 See Chapter 3, at 94.
73 This is the sense in which Bertrand and Russell put forward their formal system as a foundation for mathematics. If one accepts the axioms of the systems as true, and the rules of the game as valid, it follows as a logical consequence that any results of playing the game are also true statements.

The platonist account, which build mathematics upon the existence of external, abstract objects which exist independently of humanity, does not on the face of it depend on thought. The independence of these objects means that they would exist whether humans think about them or not. However, thought remains at the centre of the platonist accounts, either as a means of access of the objects in question through some form of mental reflection or intuition.[74] At the very least, the independence of these objects of sensory experience is not inconsistent with the mental nature of mathematical activity noted at the outset of this section.[75]

Social constructivism is, at least on the face of it, harder to reconcile with a thought requirement, to the extent that it focuses attention on the social construction of mathematics through "social acts, decisions, or practices"[76] of mathematicians. Hersh for example seems to deny the objects of mathematical practice are abstract entities, being neither mental nor physical.[77] For this reason, it may seem convenient to play down the role of thought in favour of a focus on expression in the social constructivist account, which seems to accord with the importance attached to act of sharing.[78] Apart from the mental nature of many of the practices of individual mathematicians noted above, on a social constructivist account the ability of the individual mathematical practices to centralise "around a (coherent) characterization of the structure of the domain in question"[79] follows from "the objective nature of the logical tools used in the characterization and constitution of [mathematics]".[80] So logic, and therefore thought, lie at the heart of the structuralist account. Further, the introduction of new mathematical theories is a creative endeavour,

---

74 Whilst the previous chapter noted the vagaries of how the realm of platonic objects is accessed, Plato believed that these objects, "can only be seen with the eye of the mind": Plato, "The Republic" in *The Dialogues of Plato translated into English with Analyses and Introductions by B. Jowett, M.A. in Five Volumes* (3rd ed, Oxford University Press, 1892) at line 511 (Stephanus numbering). Elsewhere, Plato asserts that this realm of objects is remembered from a past life: see Shapiro at 52. See also Kurt Gödel, "Russell's Mathematical Logic" in 1944 in Paul Benacerraf and Hilary Putnam, *Philosophy of Mathematics* (2nd ed, Cambridge University Press, 1983) at 449, discussed in Stewart Shapiro, above n66 at 205.

75 Plato in particular placed mathematics at the centre of his philosophy, noting that it "naturally awakens the power of thought ... to draw us towards reality": Plato *The Republic*, at 521, cited in Shapiro, above n66 at 62. Plato, in distinguishing the role of sensory experience notes that *"the real object of the entire subject is ... knowledge ... of what eternally exists, not of anything that comes to be this or that at some time and ceases to be"*: at 527a, cited in Shapiro at 56 (emphasis added).

76 Julian C. Cole, "Mathematical Domains: Social Constructs?" in Bonnie Gold and Roger Simons (eds), *Proof and Other Dilemmas: Mathematics and Philosophy* (Mathematics Association of America, 2008) at 113.

77 See Cole, above n76 at 119–120.

78 See below.

79 Cole, above n76 at 117.

80 Cole, above n76 at 118.

the act of creation being a primarily mental activity.[81] Similarly, Field's fictionalist account, which asserts that mathematics is no more than a useful short-cut adopted by convention, implicitly acknowledges that, like any work of fiction, mathematics is a human (and therefore mental) creation.

Perhaps the philosophy least inclined to the characterisation of mathematics might be thought to be empiricism. Yet Shapiro notes that

> [s]ince mathematical knowledge seems to be based on proof, not observation, mathematics is an apparent counterexample to the main empiricist thesis. Indeed, mathematics is sometimes held up as a paradigm of *a priori* knowledge – knowledge prior to, and independent of, experience. Virtually every empiricist took the challenge of mathematics most seriously, and some of them went to great lengths to accommodate mathematics, sometimes distorting it beyond recognition (see Parsons [Mathematics in Philosophy] 1983: Essay 1).[82]

The importance of thought to mathematics is abundantly clear in Quine's quasi-empiricist philosophy, in which mathematics forms part of a web of *belief* which is validated by empirical observation, rather than forming the subject of any direct empirical observation. Such a position is consistent with the traditional empirical view, under which mathematics could be characterised as a system of beliefs which ultimately derive their truth value from empirical observation.

It is submitted that the close relationship between mathematics and philosophy evident in the philosophical debates can also be understood independently as further evidence of the nature of mathematics as a mental activity. Apart from the philosophical accounts of mathematics just discussed, the content of mathematics itself at times raises philosophical questions. For example, the Skolem paradox raises issues about "the human ability to characterize and communicate concepts";[83] Cantor's continuum theory may have implications for our understanding the nature of truth and proof;[84] and Gödel's incompleteness theorem has implications for our understanding of the way the human mind operates.[85]

4.2 Expression in mathematics

A *Expression in the philosophies of mathematics*

Some philosophies of mathematics clearly embrace expression as a key component of mathematics. In particular, nominalist accounts such as fictionalism

81 Cole, above n76 at 121, citing Hersh with approval. The nature of mathematics as a creative endeavour is touched upon in Section 4.2.
82 Shapiro, above n66 at 3.
83 Shapiro, above n66 at 40.
84 See the discussion in Shapiro, above n66 at 42–43.
85 The incompleteness theorem is summarised below. On the implications of the theorem for the nature of the mind, see Shapiro, above n66 at 43–44. The implications for such a conclusion for our understanding of software is discussed further in the next chapter.

and social constructivism emphasise the nature of mathematics as a cultural institution. On these accounts, it is the sharing of mathematics, and the shared acceptance which that engenders, which justifies the truth of mathematical propositions. That is, in order for 'new' mathematical theories to become knowledge, their truth must be ascertained by reference to the community of mathematicians who review and criticise them. This ability to share obviously requires that these propositions be expressed in some form. As such it is entirely consistent with the 'marketplace of ideas' theory of free speech.[86]

Similarly, for the formalist account, the expression of mathematics in symbolic form on paper is the ultimate manifestation of mathematics. To understand how formalism fits with free expression we have only to understand what the individual deductive steps in a formal system represent. Axioms are fundamental assumptions so evident that they can be taken at face value. Theorems are derived from axioms by a set of allowed deductions. A deduction is "a process of reasoning by which a specific conclusion necessarily follows from a set of general premises".[87] In other words, a deductive step can be thought of as a fundamental discrete unit of logical thought. The purpose of formal systems is to establish the truth or falsity of mathematical hypotheses according to a strict set of rules. The importance of the formalist enterprise is that it is intended to be communicated to other mathematicians in order that the claimed 'proof' can be accepted. It is clear therefore that the formalist conception of mathematics depends on the free transmission of formal proofs, and therefore freedom of expression is required.

Intuitionism takes what is perhaps the most contrary view, in that the assertion that "[i]n its simplest form mathematics remains confined to one mind."[88] It was noted by Heyting in the discussion in Chapter 3 that mathematics "cannot be rendered exactly by means of language".[89] Yet the intuitionist account does not preclude expression, merely highlights the difficulties attendant on communicating mathematics between individuals. Whilst for an intuitionist, "[m]athematical ideas are independent of the dress of language",[90] the practice of mathematics, even for intuitionists, necessarily involves such expression.

This concern with the mental aspects of mathematics is also evident in the platonist account. As was noted above. Given the interplay between expression and thought noted by Vygotsky above,[91] it is clear that in any event, the need to

---

86 See *Abrams v United States*, 250 U.S. 616 (1919) per Holmes J (dissenting).

87 "Deduction" in *Collins English Dictionary and Thesaurus* (Harper Collins, 1993).

88 Arund Heyting, "Intuitionistic views on the nature of mathematics," (1974) 27(1) *Synthese* 79 at 80.

89 Heyting, above n88 at 89. See also Morris Kline, *Mathematical Thought from Ancient to Modern Times* (Oxford University Press, 1972) at 1201.

90 Kline, above n89 at 1200–1201.

91 See the discussion on page 127 above. In essence, Vygotsky and Kozulin note that "[i]t is unsatisfactory and even painful to the thinker himself, if he is not permitted to communicate his thoughts to others, and it is obviously of no value to his neighbours. Moreover it is extremely difficult to hide thoughts that have any power over the mind.": Vygotsky and Kozulin, above n21.

protect freedom of thought would extend to the protection of freedom of expression, on either an intuitionistic or platonist account. This is especially the case where the net result of allowing monopolies on these ideas would conflict directly with the notion of autonomy.

Other philosophical accounts are generally silent on the role of expression in mathematics.[92] In support of the contention that mathematics requires freedom of expression then, it is apposite to note first the interconnectedness of thought and expression noted earlier in this chapter. Further it is submitted that the consistency of these accounts with mathematics can be justified through an exposition of the role of expression in mathematical practice.

It will be recalled that rationalists believed that truth could be accessed directly by human reason. The logicist school accords with this view, but defines human reason more narrowly as logic. It will also be recalled that truth is one of the justifications put forward for free speech. In both contexts, it is the application of human reason to presented facts which allows for their truth value to be assessed. Thus the rationalist understanding of maths is clearly consistent with a mathematics exception based on freedom of speech.

Empiricists believed that mathematics is like any other science, in that it ultimately depends for its validation on evidence that is observable by the senses. In this sense it stands in direct contrast to the rationalist school, so cannot be reconciled with freedom on the same basis. Yet a variation on the above argument used for rationalism is appropriate here too, as empiricism also sees mathematics as a quest for the truth. Although it is not the process of reason by which truth is assessed, it is still to be determined according to objective criteria. Mathematical truths, once proved, then are much like laws of nature in as much as they are a correct explanation of a set of evidence.

On the empiricist approach, a mathematical innovation, advance or hypothesis, is to be accorded no weight until proved true. Up until this point then, it is nothing more than an idea. There is also a fundamentally democratic understanding shared between the two that any new theory should be freely put forward, and then tested for its truth value which mathematics then shares with science.[93] This is something that is understood in mathematics, as any grand theory, such as Hilbert's formalist approach, was open to be knocked down as it was by Gödel, regardless of which mathematician had the better reputation at the time.

B *Mathematics as a language*

To the non-mathematician, mathematics is a swirling morass of symbols. Further, the understanding of mathematics as dominated by rigour and directed

---

92 The remaining accounts are platonism, rationalism, empiricism, transcendental idealism, logicism, quasi-empiricism, and those structuralist accounts which depend on the aforementioned philosophies in determining the nature of the structures involved.

93 See Robert K. Merton, "Science and Technology in a Democratic Order" (1942) 1 *Journal of Legal and Political Sociology* 115.

to issues of proof might suggest no room for expressivity. Thus to demonstrate the expressive nature of mathematics as a language, the similarity between mathematics and literature will be illustrated. Consider the following excerpt translation of the Russian novel *Eugene Onegin* by Aleksandr Pushkin, as translated by James Falen:[94]

> And then with verse of quickened sadness
> He honored too, in tears and pain,
> His parents' dust... their memory's gladness...
> Alas! Upon life's furrowed plain —
> A harvest brief, each generation,
> By fate's mysterious dispensation,
> Arises, ripens, and must fall;
> Then others too must heed the call.
> For thus our giddy race gains power:
> It waxes, stirs, turns seething wave,
> Then crowds its forebears toward the grave.
> And we as well shall face that hour
> When one fine day our grandsons true
> Straight out of life will crowd us too!

Now consider a translation of the same passage by a different translator, Douglas Hofstadter:[95]

> And there he, on the the stark, dark marker
> Atop his parents' graves, shed tears,
> And praised their ashes – darker, starker.
> Alas, life reaps too fast its years;
> All flesh is grass. Each generation,
> At heaven's hidden motivation,
> Arises, blooms, and falls from grace;
> Another quickly takes its place.
> And thus our race, rash and impetuous,
> Ascends and has its day, then raves
> And hastens toward ancestral graves.
> All too soon, death's sting will get to us;
> Aye, how our children's children rush
> And push us from this world's sweet crush.

A side-by-side comparison of these examples makes clear both the similarity of the underlying concepts, and the way in which the translators have injected their own individual style into the translation. From the copyright perspective, this example aptly illustrates differences between idea and expression, and that the

---

94 See Tal Cohen, "An Interview with Douglas Hofstadter," 11 June 2008, <http://tal.forum2.org/hofstadter_interview> (15 June 2008).
95 Taken from Cohen, above n94.

two are sufficiently different to warrant individual protection. Similar patterns emerge in mathematics. For example, the following is an excerpt of a famous mathematical proof, Gödel's Incompleteness Theorem, which was introduced briefly in Chapter 1, as it appeared in an English translation of Gödel's 1944 paper:

> **Proposition VI**: To every w-consistent recursive class c of *formulae* there correspond recursive class-signs r, such that neither v Gen r nor Neg (v Gen r) belongs to Flg (c) (where v is the free variable of r).[96]
>
> **Proposition XI**: If c be a given recursive, consistent class of formulae, then the propositional formula which states that c is consistent is not c-provable; in particular, the consistency of P is unprovable in P, it being assumed that P is consistent (if not, of course, every statement is provable).[97]

Whilst the ordinary reader might be able to identify these statements as mathematical in nature, they are likely to be largely incomprehensible beyond that, without prior knowledge or further explanation, much as Pushkin's *Eugene Onegin* in the original Russian would be largely incomprehensible to an English speaker. Putting that aside however, the meaning of these two theorems can be summarised as follows:

> Any sufficiently powerful formal system cannot be both complete and consistent.[98]

This statement is much more digestible, despite some remnants of mathematical jargon, but contains a similar idea to the following:

> All we know of the truth
> Is that absolute truth, such as it is,
> Is beyond our reach.[99]

On this basis it is apparent that mathematics is a form of language, as impenetrable to some as Japanese or Arabic might be to others, yet with its own peculiar pleasures:

> Mathematics is in the first place a language in which we discuss those parts of the real world which can be described by numbers or by similar relations of

96 Kurt Gödel, *On Formally Undecidable Propositions of Principia Mathematica and Related Systems* (B. Melzer (trans), Dover Publications, 1992) at 57.

97 Gödel, above n96 at 70 (footnotes omitted).

98 The above simplification of the theorem is in the author's words, but is is based on a number of different formulations, influenced by the following sources: Douglas Hofstadter, *Godel, Escher Bach: An Eternal Golden Braid* (Basic Books, 1980); "Formalism" on *Bookrags.com* <http://www.bookrags.com/research/formalism-wom/> (14 July 2007).

99 Nicholas of Cusa (1401–64) De Docta Ignorantia. Cited in Paola Zizzi, "Poetry of a Logical Truth" <http://www.thalesandfriends.org/en/papers/pdf/zizzi_paper.pdf> (16 June 2008).

order. But with the workaday business of translating the facts onto this language there naturally goes, in those who are good at it, a pleasure in the activity itself. They find the language richer than its bare content; what is translated comes to mean less to them than the logic and the style of saying it; and from these overtones grows mathematics as a literature in its own right. Mathematics in this sense is a form of poetry, which has the same relation to the prose of practical mathematics as poetry has to prose in any other language. The element of, the delight in exploring the medium for its own sake, is an essential ingredient in the creative process.[100]

Once it is accepted that mathematics is capable of expressing ideas, it might be wondered why the language of mathematics is so unlike natural language. The reason is clear from the following passage:

> The main reason for using mathematical grammar is that the statements of mathematics are supposed to be completely precise, and it is not possible to achieve complete precision unless the language one uses is free of many of the vaguenesses and ambiguities of ordinary speech. Mathematical sentences can also be highly complex; if the parts that made them up were not clear and simple, then the unclarities would rapidly accumulate and render the sentences unintelligible.[101]

This might be little more than a quaint observation were it not for the importance which a recharacterisation of a mathematical problem can have. In many circumstances, the way in which a problem is expressed can lead to solutions, insights, or connections with other areas of mathematics.[102] It is this fluidity which is required for the advance of mathematics, and which directly demonstrates both the need for freedom of expression, and the interrelationship between expression and thought.

C   Mathematics as an aesthetic activity

This interplay between expression and thought hints at the inadequacy of a teleological understanding of mathematics, and directs attention towards the motivational factors which drive the mathematical enterprise forward. Although the usefulness of mathematics to the sciences is not disputed, many mathematicians have describe their discipline by reference to the concept of beauty.

The interrelationship of mathematics and beauty was introduced by the Pythagoreans, for whom number was the essence of everything,[103] and for

---

100 Jacob Bronowski, *Science and Human Values* (Pelican, 1964) at 21.
101 "I.2 The Language and Grammar of Mathematics" in Gowers et al. (eds), above n57.
102 See generally "I.4 General Goals of Mathematical Research" in Gowers et al. (eds), above n57.
103 Mark Steiner, "Mathematics Applied: The Case of Addition" in Gold and Simons (eds), above n76 at 321. For a modern version of this position see Steven Weinberg,

whom beauty corresponded with symmetry.[104] Similarly, Aristotle noted that "[t]hose who assert that the mathematical sciences say nothing of the beautiful are in error. The chief forms of beauty are order, commensurability, and precision."[105] Thus for Aristotle, it was only by reference to mathematics that beauty could be defined.[106]

More recently, Bertrand Russell, whose practical influence on both mathematics and the philosophy of mathematics was discussed in Chapter 5, described the beauty of mathematics as follows:

> Mathematics, rightly viewed, possesses not only truth, but supreme beauty – a beauty cold and austere, like that of sculpture, without appeal to any part of our weaker nature, without the gorgeous trappings of painting or music, yet sublimely pure, and capable of a stern perfection such as only the greatest art can show.[107]

Similarly, Hardy noted that "[t]he mathematician's patterns, like the painter's or the poet's, must be *beautiful*; the ideas, like the colours or the words, must fit together in a harmonious way. Beauty is the first test: there is no permanent place in the world for ugly mathematics."[108] The eminent French mathematician, Henri Poincaré similarly drew attention away from the useful aspects of mathematics towards the beautiful by noting beauty as a motivational factor for mathematicians:

> The mathematician does not study pure mathematics because it is useful; he studies it because he delights in it and he delights in it because it is beautiful.[109]

Whilst these observations of a few mathematicians might be dismissed as elitist or frivolous, and of limited applicability to the broad practice of mathematics, they are but a few examples of a "long tradition in mathematics of describing proofs and theorems in aesthetic terms",[110] and other claims by mathematicians that

---

"Towards the final laws of physics," in *Elementary particles and the laws of physics: the 1986 Dirac Memorial Lectures* (Cambridge University Press, 1989), cited in Steiner, *ibid.*

104 Nathalie Sinclair, "The Roles of the Aesthetic in Mathematical Inquiry" (2004) 6(3) *Mathematical Thinking and Learning* 261 at 263.

105 Aristotle, *The Metaphysics of Aristotle* (H.G. Bohn, 1857) XIII 3.107b.

106 In recent times, Schmidhuber has advanced an algorithmic theory of beauty: Jürgen Schmidhuber, "Low Complexity Art" (1997) 30(2) *Leonardo* 97.

107 Bertrand Russell, "The Study of Mathematics" in Andrew Brink, Margaret Moran and Richard A. Rempel (eds), *Collected Papers of Bertrand Russell, Volume 12* (Routledge, 1985) 83 at 86.

108 Godfrey H. Hardy, *A Mathematician's Apology* (Cambridge University Press, 1992) at 85.

109 Jules Henri Poincaré, "Chapter 1: The Selection of Facts" *Science et méthode* (Francis Maitland (trans), 1908), at 22.

110 Nathalie Sinclair, "Aesthetics as a liberating force in mathematics education?" (2009) 41 *ZDM Mathematics Education* 45, at 45.

"their subject is more akin to an art than it is to a science, and, like the arts, [having] aesthetic goals".[111]

It should also be noted at this point that what the focus is not beauty as "an objective mode of judgment used to distinguish 'good' from 'not-so-good' mathematical entities"[112] but looking instead at "process-oriented, personal, psychological, cognitive and even sociocultural roles that the aesthetic plays in the development of mathematical knowledge".[113]

With this in mind, it is possible to turn to specific examples of what is considered beautiful in mathematics. Whilst it is acknowledged that "[i]t may be very hard to *define* mathematical beauty, but that is just as true of beauty of any kind – we may not know quite what we mean by a beautiful poem, but that does not prevent us from recognizing one when we read it",[114] In order to get some sense of what mathematicians might find beautiful, consider the following example, known as Euler's identity:

$$e^{i\pi} + 1 = 0$$

In fact, it was ranked first by readers of a mathematics journal as *the* most beautiful mathematical theorem.[115] The reason for its asserted beauty is the unexpected relationship it documents between five fundamental mathematical constants:

0 – the additive identity;
1 – the multiplicative identity;
π – the circular constant;
e – the base of the natural logarithms; and
i – the imaginary unit.

That these five superstar numbers should be related in so simple a manner is truly astonishing. That Euler *recognized* such a relationship is a tribute to his mathematical power.[116]

With some notion of what mathematicians consider beautiful,[117] it is then possible to set out the ways in which these aesthetics influence mathematical practice.

---

111 Sinclair, above n110 at 45.
112 Sinclair, above n104 at 262.
113 Sinclair, above n104 at 262.
114 Hardy, above n108 at 85. See also the discussion of the fine arts below, section A on page 147.
115 See David Wells, "Are these the most beautiful?" (1990) 12(3) *The Mathematical Intelligencer* 37. Eminent physicist Richard Feynman also called this "the most remarkable formula in mathematics": "Chapter 22: Algebra" *The Feynman Lectures on Physics: Volume I* (1970) at 10.
116 William W. Dunham, *Euler: the master of us all* (The Mathematical Association of America, 1999) at 98.
117 For a review of the many and varied conceptions of the mathematical aesthetic, see Sinclair, above n110 at 46–49.

The discussion in the previous chapter highlights the way in which the nature of truth is sometimes seen as the key issue for the philosophy of mathematics. However, in relation to understanding the practice of mathematics (or requirements thereof), its role is limited. This is because although multitudes of theorems are proved each year, not all of them are considered significant, in that they are a useful addition to the sum of mathematical knowledge.[118] In these circumstances, other criteria must operate to filter out what is considered valuable (or likely to be valuable), and it is these criteria which are encapsulated in mathematicians' claims to beauty or elegance.

Sinclair notes three ways in which the aesthetic operates in mathematical inquiry and influences the way in which mathematics is created: in evaluation, generation, and motivation. The evaluative aesthetic is "the most recognized and public of the three roles",[119] and concerns "judgments about the beauty, elegance and significance of entities such as proofs and theorems".[120] The evaluative role of the aesthetic in mathematics plays two roles in the practice of mathematics:

> First, it mediates a shared set of values amongst mathematicians about which results are important enough to be retained and fortified. ... Second, the aesthetic determines the personal decisions that a mathematician makes about which results are meaningful, that is, which meet the specific qualities of mathematical ideas the mathematician values and seeks.[121]

The generative role of the aesthetic in mathematics involves "free, orderly, aesthetic exploration"[122] of a problem domain, looking to generate "new ideas and insights that could not be derived by logical steps alone".[123] Sinclair notes the difficulty of describing the role aesthetic in the generation of mathematics, since it often operates "at a tacit or even subconscious level, and intertwined as it frequently is with intuitive modes".[124] Nonetheless, this generative role is evident in the accounts of successful mathematicians.[125]

Mathematicians are also specifically motivated to work in particular areas by aesthetic considerations, such as visual appeal, a sense of surprise and paradox, and social aesthetics.[126] Aesthetic considerations have both a selective function,

---

118 Sinclair, above n104 at 265.
119 Sinclair, above n104 at 264.
120 Sinclair, above n104 at 264.
121 Sinclair, above n104 at 267.
122 Sinclair, above n104 at 272.
123 Sinclair, above n104 at 264.
124 Sinclair, above n104 at 270.
125 See in particular, Henri Poincaré, "Mathematical creation" in James R. Newman (ed), *The world of mathematics* (Simon & Schuster, 1956) at 2041, cited in Sinclair, above n104 at 270. The importance of such indirect approaches to gaining insight is also particularly evident in the accounts of Sir Michael Atiyah, Béla Bollobás, and Alain Connes in "VIII.6 Advice to a Young Mathematician" in Gowers et al. (eds), above n57.
126 Sinclair, above n104 at 275.

assisting the mathematician to determine which areas to become involved in,[127] and also a heuristic function, in that they influences "the discernment of features in a situation, and thereby directing the thought patterns of the inquirer".[128]

5   Freedom, the useful arts, and non-patentability

It has been demonstrated how mathematics requires freedom of thought and freedom of expression. How then to give effect to such a requirement within patent law (or alternatively, to suggest how such a requirement is already a part of patent law)? The difficulties in reconciling the legal status of mathematics in patent law with the nature of mathematics as espoused in the philosophies of mathematics was touched upon in Chapter 3.

Mathematics is a logical tool often used for specific functional purposes, such as in the sciences to describe artefacts of the natural world, or as a means of expressing useful relationships. Yet mathematics is also an intangible, aesthetic, and expressive discipline. It is these characteristics which explain why mathematics has been excepted from patentability.

It is suggested that the best vehicle for importing freedom into patent law considerations is the distinction between the fine arts and the useful arts. In other words, that mathematics is predominantly a fine art, rather than a useful art.

5.1   Defining the fine arts and useful arts

To understand the distinction between the fine and useful arts, it is first necessary to explore what such terms mean. The clues to that meaning are largely historical and functional, rather than textual, although looking at the definitions of those words can assist in reinforcing those clues.

A   *What are the fine arts?*

Historically, the traditional fine arts included painting, sculpture, drawing, and architecture, although the *Encyclopedia Britannica* provides a broader listing:

> Traditional categories in the [fine] arts include literature (including poetry, drama, story, and so on), the visual arts (painting, drawing, sculpture, etc.), the graphic arts (painting, drawing, design, and other forms expressed on flat surfaces), the plastic arts (sculpture, modeling), the decorative arts (enamelwork, furniture design, mosaic, etc.), the performing arts (theatre, dance,

---

127 "A mathematician has a great variety of fields to choose from, differing from one another widely in character, style, aims, and influence; and within each field, a great variety of problems and phenomena. Thus, mathematicians must select in terms of the research they pursue, the classes they teach, and the canon they pass on.": Sinclair, above n104 at 274.
128 Sinclair, above n104 at 277.

music), music (as composition), and architecture (often including interior design).[129]

What such endeavours have in common is that they "seek expression through beautiful or significant modes".[130] From this definition it is possible to unpack two aspects, the first being the expressive nature of such arts, the second being their reliance on beauty.[131] The importance given to expression underlies the fact that the value which is to be attributed to the arts does not lie in utilitarian assessments of the functional work of the created objects. Rather, the fine arts are concerned with the "creation of objects and presentations of imagination for their own sake without relation to utility".[132]

There is no doubt that some of these arts, most obviously architecture, can contain functional elements, although it is the beautiful, or aesthetic qualities which are the predominant focus.[133] Reliance on beauty may be seen as problematic by some. Socrates for example said that "all beauty is difficult".[134] More recently, it has been pointed out that

> terms such as beautiful and ugly seem too vague in their application and too subjective in their meaning to divide the world successfully into those things that do, and those that do not, exemplify them. Almost anything might be seen as beautiful by someone or from some point of view; and different

129 Encyclopaedia Britannica, "the arts" *Encyclopaedia Britannica Online* <http://www.britannica.com/EBchecked/topic/36405/the-arts> (10 August 2011).

130 "Fine arts," *The Macquarie Dictionary*, (4th ed, 2005). Other definitions also note the importance of beauty, or aesthetics. "Art that is produced more for beauty or spiritual significance than for physical utility.": "fine arts," Eric D. Hirsch, Joseph F. Kett and James S. Trefil (eds), *The New Dictionary of Cultural Literacy* (Houghton Mifflin, 2002). The importance of beauty is also reflected in the alternative names given to the fine arts, namely the belle arts or beaux arts.

131 The word significant will be purposefully avoided, as its meaning in the artistic context is different from the way it is commonly used in law, and may cause misunderstanding. However, it is apposite to note some expanded definitions: "the creation of works of beauty or *other special significance*" (emphasis added): "art," in *Collins English Dictionary* (Collins, 2000); "Art that is produced more for beauty or *spiritual* significance than for physical utility." (emphasis added): "fine arts," in Hirsch et al. (eds), above n130.

132 Robert I. Coulter, "The Field of the Statutory Useful Arts: Part II" (1952) 34 *Journal of the Patent Office Society* 487 at 494. This point of contrast between fine and useful arts is explored further below.

133 "Arts judged predominantly in aesthetic rather than functional terms, for example painting, sculpture, and print making. Architecture is also classified as one of the fine arts, though here the functional element is also important." "fine arts" in *The Hutchinson Unabridged Encyclopedia including Atlas*, (Helicon, 2005). It has also been noted that since the mid-18th century, "the concept of 'utilitarianism' (functionality or usefulness) was used to distinguish the more noble 'fine arts' (art for art's sake), like painting and sculpture, from the lesser forms of 'applied art', such as crafts and commercial design work, and the ornamental 'decorative arts', like textile design and interior design.": "Definition of Art" <http://www.visual-arts-cork.com/art-definition> (25 March 2011).

134 Howard Caygill, *A Kant Dictionary* (Blackwell Reference, 1995) at 91.

people apply the word to quite disparate objects for reasons that often seem to have little or nothing in common.[135]

Even in the visual arts, in the wake of the cubist and dadaist movements, the traditional notion of beauty as the aim of art has come under prolonged attack, to the point where "[b]eauty had disappeared not only from the advanced art of the 1960's but from the advanced philosophy of art of that decade as well."[136] Still, the concept has its uses. Firstly, as noted above,[137] it is by reference to the concept of beauty that mathematicians have often described their work. Secondly, an understanding of artistic works as beautiful draws attention to the traditional fine arts concern with emotional responses. From this, it is possible to broaden the label to the notion of the aesthetic, which encompasses not only beauty.

The liberal arts: Although the distinction in Australia is drawn between fine arts and useful arts,[138] in the US the nomenclature 'liberal arts' is often used instead, and the definitions of the two terms suggest some difference in focus. Coulter also adds a third category which lie beyond the useful arts, which includes, business teaching and politics, referring collectively to the three groups as the 'cultural arts'.[139] Even allowing for a significant overlap between the three categories, this might be thought to create a source of ambiguity. It may also be an unimportant point, to the extent that the focus for the purposes of determining the boundaries of patent law is on whether something is a useful art or not. To that end, the description "fine arts" is probably better understood in the broad sense, that is, as all which is not a useful art. However, for the purposes of completeness, the content of the liberal arts is set out below.

The origin of the term 'liberal arts' is an interesting topic in its own right, and has much in common with the history of mathematics detailed in the last chapter. The term can be traced back to antiquity, and in particular Plato, for whom they represented a path of learning to be traversed by his Guardians in order to prepare for the study of philosophy and develop the wisdom required to rule over society.[140] Plato's scheme came to form the basis of the teachings of medieval universities. The word liberal is important in that it denotes the higher nature of the learning involved, the liberal arts forming the subject of "aristocratic pursuits and skills, as opposed to 'the mechanical arts'".[141] Put another

---

135 "Aesthetics" in *Encyclopædia Britannica* (Encyclopædia Britannica, 2011).
136 Arthur Danto, *The Abuse of Beauty: Aesthetics and the Concept of Art* (Open Court, 2003) at 25. On the influence of non-Euclidean geometries on cubists, see Tony Robbin, *Shadows of reality: the fourth dimension in relativity, cubism, and modern thought* (Yale University Press, 2006), especially Chapter 3.
137 See Section C on page 149.
138 *National Research Development Corporation v Commissioner of Patents* (1959) 102 CLR 252 ("*NRDC*") at 275.
139 See Coulter, above n132.
140 Plato, above n74, Book VII.
141 "Liberal Arts and Liberal Sciences" in McLeish (ed), above n66.

way, the liberal arts "serve the purpose of training the free man, in contrast with the [mechanical or useful arts], which are pursued for economic purposes".[142]

The seven liberal arts are traditionally broken down into two groups, the foundational *trivium* of grammar, logic, rhetoric, and the higher *quadrivium* of arithmetic, geometry, music, and astronomy. The former is concerned with the "thorough use of language in a scientific and logical way",[143] whilst the latter is concerned primarily with mathematical subject matter, and is directed to training the mind. Although the mathematical nature of arithmetic and geometry is obvious, music and astronomy were considered to be the application of mathematical concepts to real world phenomena.[144]

The notion of the liberal arts as concerned with 'higher learning' is consistent with an intuitive understanding of the nature of the fine arts. The substantial overlap is demonstrated by reference to Kant's categorisation of the fine arts according to

> the three ways in which human beings communicate with each other: through speech, gesture and tone. The arts of speech are rhetoric and poetry, those of gesture (or the 'formative arts') include the plastic arts of architecture and sculpture and the art of painting, while the tonal arts include those of music and colour. He also admits of mixed arts. The key to understanding these divisions is to remember that they refer to skills or practices and not to objects.[145]

It is clear then that any definitional difficulty is due to the dual way in which the phrase "fine arts" is used. Firstly, there is the narrow sense, where it accords with Kant's formative arts. Secondly there is the broader sense which also includes the arts of speech and the tonal arts. It is this broader interpretation which is most suitable for present purposes. Kant's categorisation is also of interest in the present context since it draws attention to the communicative aspect of the fine arts. When the fine arts are seen as a communication, or expression, from artist to audience, the importance of freedom of expression is underscored.

---

142 Otto Willmann, "The Seven Liberal Arts" in *The Catholic Encyclopedia*, Volume 1 (Robert Appleton Company, 1907) <http://www.newadvent.org/cathen/01760a.htm> (31 Mar 2011). See also Caygill, above n134 at 85.

143 "A reading list in the 'classics,' from ancient to modern" <http://triviumquadrivium.wordpress.com/2010/02/13/is-it-worth-reading-the-classics-if-so-what-should-i-read/> (30 March 2011).

144 This reflects the Pythagorean influence "that mathematics offered a key to understanding reality, whether this reality was conceived to have an underlying geometrical structure … or whether it was simply seen as ordered and 'in proportion.'": Serafina Cuomo, "VI.1 Pythagoras" in Gowers et al. (eds), above n57. Astronomy was concerned with building a "geometrical model of the cosmos": Ian Mueller, "MATHEMATICS, Earlier Greek" in Donald J. Zeyl, Daniel Devereux and Phillip Mitsis (eds), *Encyclopedia of Classical Philosophy* (Greenwood Press, 1997) and music, or harmonics, was concerned with such things as "numerical expression of the fundamental musical concords": Mueller, *ibid*.

145 Caygill, above n134 at 86.

B   What are the useful arts?

As with fine arts, it is important as a starting point to define what is meant by the "useful arts". In the US the Constitution allows Congress to pass laws to promote the progress of useful arts.[146] It is clear then that in this jurisdiction, any attempt to go beyond the bounds of "useful arts" is invalid because it is unconstitutional – a strong argument against patentability indeed.[147] A contrast between the non-patentable fine and patentable useful arts was also drawn in *NRDC*. It will be shown how the definition of this term has a clear influence on the scope of patent law in Europe as well.

In 1952, Coulter provided what is "still the most exhaustive and deeply considered"[148] attempt at defining the useful arts in the patent law context.[149] Coulter's analysis depended on both an exploration of the classical notion of the useful arts, as well as consideration of the intention of the Framers of the US Constitution.

It has already been noted that the useful arts were typically contrasted with the fine arts, or the liberal arts. The contrast with the liberal arts is particularly noteworthy, since it dates back to classical times when the liberal arts were the areas of study reserved to free men, whereas the useful, or manual, arts, were those works carried out by slaves. This distinction was carried through to the Renaissance, wherein the liberal arts formed the curriculum at medieval universities – the province of the aristocracy. In contrast, the practitioners of the manual or useful arts

> did not require a high degree of intellectual attainment and cultural education, and rarely possessed them, and they engaged in manual labor, which accounted in part for their more or less lowly social and economic position in the English class structure.[150]

Built around these manual and useful arts were the guilds, through which these arts were taught by serving as apprentices.[151] It is at least arguable that this understanding of the useful arts is reflected in the *Statute of Monopolies*,[152]

---

146 "The Congress shall have the power ... [t]o promote the Progress of ... useful Arts, by securing for limited Times to ... Inventors the exclusive Right to their ... Discoveries": *Constitution of the United States of America*, art. 1, §8, cl. 8.

147 *Graham v John Deere Co* 383 US 1 (1966) at 5.

148 Alan L. Durham, "'Useful Arts' in the Information Age" (1999) *Brigham Young University Law Review* 1419 at 1437.

149 Coulter, above n132. Coulter's analysis formed part of a three-part series of articles in which he sought to criticize the failings of the mental steps doctrine. See also Robert I. Coulter, "The Field of the Statutory Useful Arts: Part I" (1952) 34 *Journal of the Patent Office Society* 426; Robert I. Coulter, "The Field of the Statutory Useful Arts: Part III" (1952) 34 *Journal of the Patent Office Society* 718.

150 Coulter, above n132 at 496.

151 Coulter, above n132 at 496.

152 *Statute of Monopolies* 1623, 21 Jac 1, c 3 (UK).

wherein the period of a patent corresponding in early times to the length of two apprenticeships.

Coulter's analysis concludes with a claim that the "fundamental attribute"[153] of the useful arts was that "[t]hey relate to controlling the forces and materials of nature and putting them to work in a practical way for utilitarian ends serving mankind's physical welfare."[154] Coulter then asserts that the phrase useful arts has a contemporary equivalent – technology.[155] Coulter's analysis seems to have been influential on judicial development of patent law at the time.[156]

Understood as a synonym for technology, it is possible to potentially unify the scope of patent law across the three jurisdictions thus far considered. It is apposite to note the requirement in Article 27 of TRIPS,[157] that members of the World Trade Organisation make patents available in all fields of *technology*, supports this unified understanding. In the EU, the link between the technological arts, and the requirement in of a technical contribution, is obvious. Even putting aside the use of the term "useful arts" in *NRDC* as noted above, it is at least arguable that the historical meaning of "manner of manufacture" in the *Statute of Monopolies* encompassed "technology" as it existed at that time. The subsequent expansion of the meaning of manner of manufacture beyond a literal interpretation might be understood as a reflection of the changing nature of technology over the subsequent period.

Defining technology: But can technology be properly defined, or is this fool's errand? Durham suggests that "the more one looks at how 'technology' has been defined by scholars, the less one is sure what it means."[158] In *CFPH's Application*,[159] Deputy Judge Prescott described the related term "technical" as "a restatement of the problem in different and more imprecise language ... It is a useful servant but a dangerous master."[160] Similarly, the courts in the US, whilst toying with a technological arts requirement, eventually resiled from such an understanding of the boundary of patent law in *Lundgren*.[161] In the

---

153 Coulter, above n132 at 498.
154 Coulter, above n132 at 498.
155 "Probably the best word in common usage today that expresses this idea is 'technology'.": Coulter, above n132 at 498.
156 See for example, *In re Musgrave* 431 F.2d 882 (1970) at 893: "All that is necessary, in our view, to make a sequence of operational steps a statutory 'process', within 35 USC §101 is that it be in the technological arts so as to be in consonance with the Constitutional purpose to promote the progress of 'useful arts.'"; *In re Waldbaum* 457 F.2d 997 (1972) at 1003: "The phrase 'technological arts,' as we have used it, is synonymous with the phrase 'useful arts' as it appears in Article I, Section 8 of the Constitution."
157 *Agreement on Trade-Related Aspects of Intellectual Property Rights*, opened for signature 15 April 1994, 1869 UNTS 299, 33 ILM 1197 (entered into force 1 January 1995).
158 Durham, above n148 at 1444.
159 *CFPH LLC's Application* [2005] EWHC 1589 (Pat) (*"CFPH"*).
160 *CFPH* at [14]. See the discussion at [11]–[14]. See also *Symbian Ltd v Comptroller-General of Patents* [2008] EWCA Civ 1066; [2009] RPC 1 (*"Symbian"*) at [30]–[32], discussed in Chapter 3 in which their Honours noted that the concept "could easily mean different things to different people": at [30].
161 *Ex parte Lundgren*. Appeal No 2003–2088 (2005).

EU, the technological contribution test has been criticised as having "all the disadvantages of the original obscure wording [of Article 52], with the added disadvantage of not even providing the actual legislative test."[162] Similarly in *Grant*, the Full Federal Court rejected the recharacterisation of the boundaries of patent law by reference to a science or technology requirement as follows:

> What is or is not to be described as science or technology may present difficult questions now, let alone in a future which is as excitingly unpredictable now as it was in 1623 or 1959, if not more so. We think that to erect a requirement that an alleged invention be within the area of science and technology would be to risk the very kind of rigidity which the High Court warned against.[163]

Despite this, it is submitted that the philosophy of technology, and the definitions it offers, are of direct benefit in this context. As will be seen below, many of the definitions put forward bear a remarkable resemblance to the ways in which the patentable subject matter has been defined. As such, they offer potentially relevant insights into the debate, whilst also serving to demonstrate that the choice between various definitions is in a real sense unavoidable. In forming and reforming the approach to determining patentable subject matter in the cases that come before them, judges are in effect making implicit decisions referable to the view they have taken of technology, whether broad or narrow. By bringing these issues to the fore, it is suggested that a coherent conception of the scope of patentable subject matter might be developed.[164]

The advantage of discussing philosophies of technology, and making an enunciated choice as to which is the appropriate one, has the advantage that it brings policy choices out into the open. When policy is swept under the carpet, it is difficult to review, and sees the patentable subject matter issue either collapse under its own weight, being reduced to nothing other than a form requirement, or sees ad hoc bouncing from under-protection to over-protection. Fear of the future development of our concept of technology stands to drive the development of an ever broader notion of what is patentable without clear conceptual guidance.

Whilst a comprehensive analysis of the multitudinous interpretations of technology is a mammoth task, a brief survey of possible alternatives is warranted.[165]

---

162 *Symbian* at [30].

163 *Grant v Commissioner of Patents* (2006) 154 FCR 62 (*"Grant"*) at 71.

164 "As the principal legal response to technological change, the regime of patents too has suffered from its inability to develop a coherent sense of its own subject matter": John R. Thomas, "The Post-Industrial Patent System," (1999) 10(3) *Fordham Intellectual Property Media and Entertainment Law Journal* 3 at 41.

165 For a comprehensive account of the philosophy of technology, see Carl Mitcham, *Thinking through Technology: The Path Between Engineering and Philosophy* (University of Chicago Press, 1994). For a survey of alternatives from the perspective of their relevance to patent law, see Alan L. Durham, "'Useful Arts' in the Information Age" (1999) *Brigham Young University Law Review* 1419; Thomas, above n164.

Building on the obvious relationship between science and technology, some have suggested technology as applied science.[166] This view finds a modicum of support in the case law which distinguishes between knowledge and its application to some end. However, such a definition may be criticised on the basis that technology, although underpinned by science, is not predicated upon it. To put it another way, "[t]echnology is purposive and tends ... to be positivist. The criterion is simply, does it work."[167]

A variant on that definition emphasises the role of technology in the production of what is useful or practical.[168] This would be most consistent with the "useful result" test of *Alappat*[169] and *State Street*,[170] and accords, literally speaking, with the phrase "useful arts". However, this is a potentially broad definition, since all human endeavours aim at some ultimate good.[171] Despite the fact that such a distinction does offer a potential basis to distinguish the useful from the fine arts in that the former are created for a purpose, whilst the latter are ends in themselves.[172] The danger is that such a distinction will be overlooked, a concern which is justified by the broad reach of patent law in the wake of *State Street*.[173]

A broad understanding put forward of the nature of technology is that which describes technology as concerned with "human work",[174] or "the systematic treatment of any thing or subject".[175] Such definitions flow from the central purpose which the design process plays in the development of technologies,[176] design being "subject to rational scrutiny but in which creativity is considered to play an important role".[177] The basis of technology in rational thinking, as opposed to the "free play" of aesthetic thinking seems to offer some differentiation

---

166 Durham, above n148 at 1445.

167 Donald Cardwell, *The Norton History of Technology,* (1995) at 492–3, cited in Durham, above n148 at 1445.

168 See Frederick Ferré, *Philosophy of Technology* (Prentice Hall, 1988) at 26, who describes technology as "practical implementations of intelligence." Cited in Thomas, above n164 at 39. Discussed in Mitcham, above n165 at 156–157.

169 *In re Alappat* 33 F.3d 1526 (1994).

170 *State Street Bank & Trust Co v Signature Financial Group Inc* 149 F.3d 1368 (1998).

171 Aristotle, *Nichomachean Ethics,* (Sir David Ross (trans), Oxford University Press, 1966) at 1. Cited in Durham, above n148 at 1446.

172 Mitcham, above n165 at 156.

173 Kappos et al. note that in the wake of *State Street*, patents were issued in "architecture, athletics, insurance, painting, psychology and the law itself": David J. Kappos, John R. Thomas and Randall J. Bluestone, "A Technological Contribution Requirement for Patentable Subject Matter: Supreme Court Precedent and Policy" (2008) 6(2) *Northwestern Journal of Technology and Intellectual Property* 152, at 164.

174 Peter F. Drucker, *Technology, Management and Society* (1970) at 45–46, cited in Durham, above n148 at 1448.

175 Charles Singer et al., "Preface" in Charles Singer et al. (eds) *A History of Technology* (1954) at vii, cited in Durham, above n148 at 1449.

176 Maarten Franssen, Gert-Jan Lokhorst and Ibo van den Poel, "Philosophy of Technology," *Stanford Encyclopedia of Philosophy*, <http://plato.stanford.edu/entries/technology/> (1 May 2011) at §2.3.

177 Franssen et al., above n176 at §2.3.

between the useful and fine arts respectively. Not all scholars agree with such a distinction however, with Agassi, for example, claiming that "what we call the [fine] arts ... [are] all technology".[178]

As such, there needs to be some form of limitation on the scope of patentability. Adopting a narrower interpretation of technology would limit patent law to the "making of *physical artefacts* and the physical alteration of the environment".[179] Mitcham posits that despite the divergence of meanings attributed to technology, all conceptions all admit of a "primacy of reference to the making of material artefacts".[180] Hannay and McGinn also highlight physicality as a way of distinguishing technology from other forms of human activity:

> [T]echnology differs from other activity-forms in that the natural environment – both in respect to the meteorological and creature-related threats it poses to human survival, and the spatiotemporal obstacles it presents to human desires for communication and transport – is a factor that more powerfully and more directly conditions technology than is the case with other cultural forms, for example, religion and art.[181]

Why physicality is important: The idea that to be patentable a process must be "embodied and connected with corporeal substances"[182] extends back through patent law at least as far as *Boulton v Bull*.[183] It is the inherently complex relationship between the definition of a process and the physical world which has caused problems for this class of invention since the Court split evenly on the point in that same case. Despite the broadening conception of patentable subject matter since that time, physicality remains an important limiting device in the subject matter inquiry.

There is certainly some support for a broad view of the scope of patent law, as is evident from the following:

> Whatever may be left of the [Freeman-Walter-Abele] test, if anything, this type of physical limitations analysis seems of little value because... "after

---

178 Joseph Agassi, *Technology: Philosophical and Social Aspects* (1985) at 90. Cited in Durham, above n148 at 1449–1450. See also Durham, n208 below and accompanying text.

179 Durham, above n148 at 1447.

180 Mitcham, above n165 at 152. Mitcham's definition affords a basis for distinguishing between human making and "human doing – for example, political, moral, religious, and related activities": at 153. See also Franssen et al., above n176 at §2.3: "technology is a practice focused on the creation of artifacts and, of increasing importance, artifact-based services."

181 See also Robert E. McGinn "What is Technology?" in Paul T. Durbin, (ed.) *Research in Philosophy and Technology*, Vol. I (JAI Press, 1978) at 190, describing technology as "a form of activity that is *fabricative, material product making or object transforming*, purposive (with the general purpose of expanding the realm of the humanly possible), knowledge-based, resource-employing, methodical, embedded in a socio-cultural-environmental influence field, and informed by its practitioners' mental sets" (emphasis added).

182 *Boulton v Bull* (1795) 126 ER 651 at 667.

183 Discussed in Chapter 2.

Diehr and Alappat, the mere fact that a claimed invention involves inputting numbers, calculating numbers, outputting numbers, and storing numbers, in and of itself, would not render it nonstatutory subject matter, unless, of course, its operation does not produce 'a useful, concrete and tangible result'" [*State Street Bank & Trust v Signature Financial Group*, 149 F.3d 1368] at 1374 (quoting [*In re*] *Alappat*, 33 F.3d at 1544).[184]

This "useful result" test, which for a time was in the ascendancy in both Australia and the US,[185] does not equate usefulness with physicality. However, as Breyer J noted in *Bilski*, although the scope of patentable subject matter "is broad, it is not without limit".[186] Whilst not wholeheartedly adopting a physicality-based understanding of the reach of patent law, by endorsing the machine-or-transformation test as the sole test for patentability, a majority endorsed this as a "useful and important clue".[187] The distinction between abstract idea and patentable invention in *Mayo*[188] and *Alice*[189] also supports this approach.

In Australia, the rush toward a broad view of patentability has also peaked. It was through the reinstatement of a physicality notion that limits were applied to subject matter, through a modification of the useful result test to a "a concrete, tangible, physical, or observable effect [test]".[190] Australia has, by recognising the influence of EU and US approaches in determining patentability in Australia, endorsed a physicality approach.

In the EU, a physicality requirement is inherent in the notion of technical character or effect. This reflects the influence of German patent law, which had coalesced around a notion of technicality since the Red Dove case in 1969.[191] The most far-reaching statement of this requirement can be found in the statement of the the German Federal Patent Court that "the concept of technology (Technik) constitutes the only usable criterion for delimiting inventions against other kinds of intellectual achievements, and therefore technicity is a precondition for patentability."[192] Physicality in the specific sense in which it is

---

184 *AT&T v Excel Communications* 172 F.3d 1352 (1999) at 1359.

185 This was the approach adopted in the United States in: *In re Alappat* 33 F.3d 1526 (Fed Cir, 1994); *State Street Bank & Trust Co v Signature Financial Group Inc* 149 F.3d 1368 (Fed Cir, 1998); and *Ex parte Lundgren* Appeal No 2003–2088 (BPAI, 2005). This approach was endorsed in Australia in: *IBM v Commissioner of Patents* (1991) 33 FCR 218; *CCOM Pty Ltd v Jiejing Pty Ltd* (1994) 51 FCR 260 and *Welcome Real-Time SA v Catuity Inc and Ors* (2001) 113 FCR 110.

186 *Bilski v Kappos* 130 S. Ct. 3218 (2010) ("*Bilski*") at 3258.

187 *Bilski* at 3258 per Breyer J.

188 *Mayo Collaborative Services v Prometheus Labs, Inc* 132 S. Ct. 1289 (2012).

189 *Alice Corp v CLS Bank International* 134 S. Ct. 2347 (2014)

190 *Grant* at 70.

191 BGH GRUR 1696, 672 "Rote Taube".

192 BPatG Fehlersuche 2000–07–28: Patentansprüche auf "Computerprogrammprodukt" etc unzulässig, BPatG /17W(pat)69/98, <http://swpat.ffii.org/analysis/trips/index.en.html> (11 October 2006).

meant here then means "solving problems by utilising natural forces".[193] The natural forces involved are gravity, electromagnetic force, strong nuclear force and weak nuclear force.[194]

It is acknowledged however, that within the framework of a physicality requirement, the dominant approach at the EPO favours physicality as a mere form requirement. Any reference to hardware, even a pen and paper[195] is sufficient for patentability in Munich. As noted in Chapter 2 however, clever drafting may get an applicant over the subject matter hurdle, but is unlikely to escape inventive step. UK courts have resisted this approach, and still require a technical (physical) contribution. That this is a physicality requirement is evident from the analysis of the two inventions considered in *Aerotel*. The first, the Aerotel invention, was patentable because of the "new physical combination of hardware"[196] featured in the claims. The Macrossan application, although claiming a method to be implemented on a computer, did not exhibit the requisite technical character because there was nothing technical (physical) claimed "beyond the mere fact of the running of a computer program".[197] Why that is insufficiently physical is addressed in detail in Chapter 5.[198]

As a matter of policy, it is submitted that a physicality requirement, whilst admittedly the most conservative option, serves an important function. On a theoretical level, a physicality requirement is consistent with the theoretical basis for awarding property rights in patent law, namely, that "to create something is to control its existence. Without control there is no possession and no ownership."[199]

On a practical level, the laws of physics and material properties usually offer an inventor a series of difficult technical obstacles to overcome which means that the inventor runs the risk of complete failure when trying to realise an invention. As opposed to going to market with a bad product, the inventor might not have a

193 See Foundation for a Free Information Infrastructure, "Regulation about the invention concept of the European patent system and its interpretation with special regard to programs for computers", <http://swpat.ffii.org/stidi/javni/index.en.html> (12 October 2006). The longhand version of this test, "plan-conformant activity of using controllable natural forces to achieve a causally overseeable success which is, without mediation by human reason, the immediate result of controllable natural forces", owes its origin to the German Federal Court's 1977 "Dispositionsprogramm" case. See BGHZiv 1977 Bd 67 p22ff; BGHZ 67, 22; Beschluss des X. Zivilsenats des BGH in der Rechtsbeschwerdesache X ZB 23/74. English translation available at <http://swpat.ffii.org/vreji/papri/bgh-dispo76/> (13 April 2007).
194 For a brief summary see Jeff Silvis and Mark Kowitt, "The Four Forces of Nature," *Ask An Astrophysicist*, <http://imagine.gsfc.nasa.gov/docs/ask_astro/answers/980127c.html> (9 May 2011).
195 T258/03 *Hitachi/Auction Method* [2004] EPOR 55 at [4.6].
196 *Aerotel Ltd. v Telco Holdings Ltd and in the matter of Macrossan's Application* [2006] EWCA Civ 1371; [2007] RPC 7 ("*Aerotel*") at [53].
197 *Aerotel* at [72].
198 See Chapter 5, Section 2.1 on page 167.
199 Gregory Stobbs, *Software Patents* (2nd ed, 2000) at 205, discussing the 'common thread' of US patentable subject matter cases based on the expression in *Chakrabarty* that "anything under the sun that is made by man" is patentable.

product to take to market at all. For example, Dratler notes the following physical limitations to be taken into account in developing a new machine:

> While a machine's design may appear operable in concept, in order to work in the real world it must successfully address such practical problems as: metal fatigue, strain, bending, stress fractures, vibration, corrosion, pollution, spalling, differential thermal expansion and contraction, unintended electrolysis, dust, dirt, friction, ablation, evaporation, deterioration of lubricants, electric arcing, unwanted generation of static or other electricity, and aging.[200]

Further examples relevant to the design of computer hardware components include issues of heat dissipation and transistor density.[201] Such hurdles mean that the developer of a new machine faces the problem that "an infinity of possible mechanical configurations meets an infinity of environmental forces; the end result of that interaction is anything but deterministic, anything but predictable."[202] Patents thus offer a form of insurance for the holder whilst they address those risks in order to bring their products to market. That is a laudable justification for the award of patent monopolies.

In contrast the developer of an intangible product, such as "pure" software, or a new method of doing business, relies only on a *model* of the real world. In building this model of the real world, it is not "necessary to fully describe and model the entire world".[203] Only those characteristics of the real world which are *relevant* to the purpose for which the software is being written are included. As such, the software developer faces only a finite number of variables (although the number of variables may in fact be quite large). "The elimination of irrelevant modelling characteristics and the inclusion of certain relevant ones constitutes the process of abstraction."[204] Because only a limited number of variables are involved,

> because every last one can be totalled up and its effects catalogued, then it follow[s] that only a limited number of possible variable interactions exist. All the possible outcomes, all the possible states an information processing machine or software program can enter, myriad though they be, are still predictable and determinable in advance.[205]

200 Jay Dratler, Jr., "Does Lord Darcy Yet Live? The Case Against Software and Business-Method Patents," (2003) 43 *Santa Clara Law Review* 823 at 854.
201 The current trend, known as Moore's law, has seen transistor density double on integrated circuits approximately every two years since 1958. See Wikipedia, "Moore's law" <http://en.wikipedia.org/wiki/Moore's_law> (6 November 2008).
202 Gary Dukarich, "Patentability of Dedicated information Processors and Infringement Protection of Inventions that Use Them" (1989) 29 *Jurimetrics Journal* 135 at 147. As such, Dukarich calls such technologies "non-deterministic".
203 Dukarich, above n202 at 141.
204 Dukarich, above n202 at 142.
205 Dukarich, above n202 at 142.

The result of this predictability is a lower degree of risk for the inventor, as acknowledged by Stevens J in *Bilski v Kappos*:

> Business innovation ... generally does not entail the same kinds of risk as does more traditional, technological innovation. It generally does not require the same "enormous costs in terms of time, research and development" *Bicron* 416 U.S. @ 480 and thus does not require the same kind of "compensation to [innovators] for their labor, toil and expense" *Seymour v Osborne* 11 Wall. 516, 533–544 (1871).[206]

The practical impact of these different risk levels are also reflected in the nature of competition in the relevant market, in the way they impact on lead time:

> Hardware products produce lead time because reverse-engineering hardware entails a substantial expenditure of time. Would-be imitators must first determine how the new product works and what manufacturing processes were used to produce it, then adapt or build the requisite manufacturing facilities, and then build and test their own products. Software is different. There are typically no manufacturing processes to analyze, and no special factories to set up. Software is written and tested; it is then published, like books, records, or videotapes. It is possible to copy a computer program in seconds and readily reproduce that copy by the hundreds or thousands. It is more difficult, but nonetheless relatively easy, to adapt, translate, or "port" a program, and thereby appropriate much of the value inherent in the original author's creation. Software, by its nature, lends itself to quick and unexpected duplication and even translation.[207]

What this highlights is the fast-paced nature of innovation in non-physical endeavours, which is hard to reconcile with the length of protection afforded by the patent grant. It also calls into question the suitability of a regime in which award of monopoly rights, at the end of a sometimes lengthy period of examination, may occur well after the shelf-life of the innovation has expired.

Physicality of the useful arts against intangibility of the fine arts: Durham raises an interesting objection to the usefulness of a physicality requirement in distinguishing between the useful and fine arts. Durham notes that

---

206 *Bilski* at 3254 (2010), citing in support: Dan Burk and Mark A. Lemley "Policy Levers in Patent Law" (2003) 89 *Vanderbilt Law Review* 1575 at 1618; Michael A. Carrier, "Unraveling the Patent-Antitrust Paradox" (2002) 150 *University of Pennsylvania Law Review* 761 at 826; David S. Olson, "Taking the Utilitarian Basis for Patent Law Seriously: The Case For Restricting Patentable Subject Matter" (2009) 82 *Temple Law Review* 181 at 231.

207 Anthony L. Clapes, Patrick Lynch and Mark R. Steinberg, "Silicon Epics and Binary Bards: Determining the Proper Scope of Copyright Protection for Computer Programs" (1987) *UCLA Law Review* 1493 at 1509.

most abstract of "arts" have their physical manifestations and effects on the material world. The field of law, for example, produces contracts and statutes written on paper, and it alters conditions and conduct in the "real world".[208]

This is true, but it is submitted that this is not a sufficient reason to abandon a physicality requirement as Durham contends. As a matter of intuition, and as a matter of analysis, it is possible to distinguish between the cultural "arts" and technological "making and using". The key word in Durham's complaint is "manifestation" – a legal agreement which manifests itself in a written contract is not patentable, because the advance over what existed before is in the agreement, a meeting of the minds, which is clearly non-physical. The issue that Durham draws attention to is one of proper characterisation of an invention. Clever claim drafting can only be properly addressed by looking to the substance rather than the form of claims.[209]

Another answer to Durham's criticism is found in Kant's classification of the fine arts discussed above.[210] It was noted that the categories of the fine arts can be classified by the way in which humans communicate. When the fine arts are seen as communications between artist and audience, then it becomes clear why the physicality of the objects produced is not the essence of the fine arts. It is the *message* communicated by the artistic work which is its essence, and which is significant. It is that message which will be evaluated in judging the worth of a particular object independent of any physical functionality – such physical functionality is secondary in importance in the fine arts. With useful arts, the significance lies in "what is achieved in physical fact",[211] the particular physical advantage which the invention offers, which will be adjudged in determining its worth.

Mitcham, whilst acknowledging that both the fine arts and useful arts involve some sort of design process to achieve an end product (which may or may not be physical), sets out in detail the difference between the two processes:

> one can ... distinguish engineering design from artistic design according to ideals or ends in view. The engineering design ideal of efficiency stands in contrast to the artistic design ideal of beauty. Beauty is not so much a question of materials and energy as of form. About this the whole subject of aesthetics has more to say, whereas it is ethics or politics that would incorporate a philosophical evaluation of efficiency.
>
> Yet the difference between these two types of design does not remain at the level of ideals; it penetrates to the design activity itself. Efficiency refers to a process – is a criterion for choosing between processes or products

---

208 Durham, above n148 at 1448.
209 To this end, the four-step approach of the Court of Appeal in *Aerotel* provides useful guidance.
210 See page 149 above.
211 *RCA Photophone* at 191.

conceived as functioning units – whereas beauty is in the primary instance a property of stable objects. ... The ends of artistic design must be formal, whereas the final causes of engineering actions are justified in terms of human needs, wants or desires.

A further observation: Engineering design limits itself to material reality (metaphysically, matter and energy are both matter as contrasted with form). This limitation is to be grasped or approached however, by means of a mathematical calculus of forces closely associated with classical physics (Galileo and Newton) and its specific mathematical abstraction. The picturing or imagining that goes on in engineering design is done, as it were, through a grid derived from this physics – the grid itself being articulated, in the first instance, as the engineering science of mechanics. This viewing of matter and energy through the grid of classical physics gives engineering design a rational character not found in art. Engineering images, unlike other images are subject to mathematical analysis and judgment; this is their unique character and one that sometimes leads people to confuse them with thinking in a deeper sense.

Art is also concerned with imagining, but its images cannot be qualitatively analyzed – they are not subject to any well-developed calculus. Thus art, in contrast to engineering, appears as both more intuitive and more dependent on the senses.[212]

This account clearly accords with the analytical framework offered above. It makes clear that engineering (or the useful arts) is both directed towards physical outcomes, and limited during the design process by the "grid of classical physics", from which Mitcham asserts that the rationality of engineering design emerges, in contrast to the intuitive and sense-dependent design in the fine arts (considered in this chapter so far under the banner of the aesthetic). It is also clear from the quoted passage that the purpose of technology, sourced as it is in physical "human needs, wants or desires"[213] is distinct from the "formal" (or expressive) ends of artistic design.

5.2 The three dimensions of the fine arts versus useful arts distinction

The discussion above leads to a distinction between the fine arts on one hand, and the useful (or technological) arts on the other according to three distinct dimensions of analysis. These are set out in Table 4.1.

First, it is asserted that the fine arts are concerned with the aesthetic whereas the useful arts are concerned with the rational. That is, they may be created in order to provoke an aesthetic reaction, or perhaps more importantly in the current context, aesthetic considerations motivate their practitioners. In contrast,

---

212 Mitcham, above n165 at 229–230.
213 Mitcham, above n165 at 229.

Table 4.1 Fine arts versus useful arts

Aesthetic	Rational
Expressive	Purposive
Intangible	Physical

technology is rational, being centred on the process of design. Its practitioners are driven by the desire to achieve efficiency gains. Second, it is contended that fine arts are expressive whereas the useful arts are purposive. They are sometimes said to be 'ends in themselves', which is to say that the final form in which they manifest themselves is secondary in importance to the expression that the represent. Technology on the other hand is purposive, being designed to solve a particular problem, to achieve a particular result. Third, it is suggested the preferable interpretation of the useful arts is that they deal with the physical. Fine arts on the other hand are said to be intangible in nature. Whilst they may (and indeed often are) manifested in a physical form, their essence, or importance, lies beyond that manifestation.

Each of these dimensions is now considered in turn.

A   Aesthetic versus rational

It has been demonstrated that mathematics is primarily a mental activity.[214] Therefore it falls to determine whether it is a mental activity which is aesthetic, in that it is based on intuitive, sense-dependent considerations, or one which is rational, in that is a rational, engineering-like activity which is directed towards the satisfaction of the physical wants, needs, and desires of humanity.

The role of logic and the apparent availability of objective criteria upon which to judge a mathematical proof weighs in favour of a view of mathematics as rational activity. Yet this objectivity is to an extent superficial and misleading. The unresolved foundational crisis in mathematics, discussed in Chapter 5, makes it clear that whilst mathematicians seek to remove any reliance on intuition, this cannot be done. Irrational numbers, imaginary numbers, and non-Euclidean geometries were all rejected as impossible and irrelevant at various stages through history, yet their usefulness to (often unexpected) real-world applications saw their eventual validity. Conjectures and hypotheses, whilst unproven, still act as a very real motivation to mathematicians, and it has been through searching for proofs that interesting mathematics has come about.

Perhaps the most telling statement of the limits of objectivity and logic in mathematics was Gödel's incompleteness theorem, by which it he demonstrated the impossibility of excluding uncertainty, in the form of incompleteness or inconsistency, from any powerful mathematical system.

214 See Section 4.1 above.

Further, the role of the aesthetic in motivating mathematicians, and generating mathematical advances was explored earlier in the chapter. Thus it seems appropriate to conclude that the balance of evidence supports the primacy of the aesthetic over the rational in the field of mathematics.

B   *Expressive versus purposive*

On any account of mathematics, it must still be acknowledged that mathematics is useful, in that it is often a tool which is used by scientists and engineers in their work. However, the role of expression in mathematics has also been documented. Mathematics is a expressed in a symbolic language, which is unlike "natural" language, but which is nonetheless expressive.

The traditional status of mathematics as one of the liberal arts curriculum also confirms this categorisation. It is obvious from the content of the enumerated liberal arts that mathematics was not only a substantial component, but also that mathematics was considered a form of higher learning, based as it was in the *Quadrivium*, rather than the *Trivium*. Therefore, on any understanding of the term liberal arts, mathematics is clearly within its ambit.

C   *Intangible versus physical*

The role of mathematics in explaining real-world phenomena in the physical sciences might be thought to lend weight to the classification of mathematics as directed towards the physical. It may have some relevance to human needs, wants, and desires to the extent that mathematical advances lead to better understanding of the physical world, and thereby to better utilisation of it. However, this is at best an indirect physical relevance.

It was established above that mathematics is primarily a mental activity, and whilst it is often expressed in a symbolic language, that physical manifestation is not the essence of mathematics, or the grid through which its development might be viewed. To the extent that mathematics is manifested in physical form, its significance lies not in the production of printed documents. Even on the formalist account of mathematics being nothing more than a game played with symbols on paper, it is clear that mathematics is the game, not the symbols.

D   *Summary*

The lesson to be learned from this understanding of mathematics, and the importance of freedom to innovation in the field, is that it is a trap to focus on the "usefulness" of mathematics, at the expense of a critical analysis of the nature of mathematical creativity – its intangible, expressive, and aesthetic characteristics. If there is to be a way forward, it must be to balance the benefit to society of encouraging inventive or creative activity against the impact of awarding monopoly rights on the interests of others practising in the field. This sort of utilitarian analysis should lie at the heart of the patentable subject matter inquiry. As touched on both in Chapter 1 and in the discussion of the formalist school in both Chapter 5 and this chapter, mathematics and computer software are iso-morphic, so it follows that software would be expected to be non-patentable as well.

6   Conclusion

This chapter has explored the role of the freedoms of thought and expression in mathematics. Through that exploration, the nature of mathematics as a primarily mental activity, expressed in a symbolic language, and driven forward by aesthetics has become apparent. It was then demonstrated how the need for such freedoms can be reconciled with an understanding of mathematics as being a fine art rather than a useful art, and hence outside the scope of patentable subject matter. In establishing the distinction between fine and useful arts, a three-dimensional analysis was adopted, and it was demonstrated how on each dimension of the analysis, mathematics falls within the fine arts.

Having thus established the role of freedom within the mathematics exception, and its proper classification outside the scope of patent law, There is an immediate flow-on consequence to the patentability of software in two respects. First, the isomorphism between mathematics and software development means that, at least as a starting point, it should be expected that software is not patentable subject matter, being a similarly creative activity. Second, it is now possible to bring a greater deal of clarity to the related field of computer software, using that understanding of the non-patentability of mathematics, and the analytical tools developed to consider if or when software should be considered patentable. It is to that task that the next chapter of this thesis is devoted.

5 Why programming is not among the useful arts

It seems beyond question that the machines – the computers – are in the techno-logical field, are part of one of our best-known technologies, and are in the "useful arts" rather than the "liberal arts," as are all other types of "business machines," regardless of the uses to which their users may put them. How can it be said that a process having no practical value other than enhancing the internal operation of those machines is not likewise in the technological or useful arts?[1]

1 Introduction

The last chapter demonstrated that the requirements of mathematical activity locate it properly outside the useful arts. It will be recalled from Chapter 1 that there is an isomorphism, or structural similarity, between the activities involved in mathematics and programming.[2] In the discussion of formalism in Chapter 3, it was briefly noted that there is a link between formalism and the development of a formal theory of computing.[3] This chapter will consider how far the isomorphism and formal connection extend the asserted basis of the non-patentability of mathematics to software.

As the previous chapter did for mathematics, the expressive, aesthetic, and intangible nature of software development will be explored. The existence of these attributes, it will be argued, supports the contention that the classification of mathematics as a fine art will usually be appropriate to software. In particular, it will be suggested that the interrelationship of software and hardware, referred to as context in Chapter 1, acts as an important consideration, which becomes more important the higher up the abstraction chain one goes. Where the soft-ware and hardware are closely interrelated, the physical limitations may constrain the software such that expressive and aesthetic considerations are similarly limited. Where software is merely ancillary to a physical device however, this

---

1 *In re Benson* 441 F.2d 682 (CCPA, 1971) at 688.
2 See Chapter 1, at 30.
3 See Chapter 3, at 94.

does not change the nature of programming, but it does in effect change what is sought to be patented. The limited scope of the claims will mean both that programming as a creative activity is not disrupted, and that what is claimed is sufficiently connected to the traditional conception of the useful arts that is not contentious. Determining whether it is software, or a physical device which is actually being claimed requires difficult determinations of fact in any particular case, but the difficulty of the choice does not obviate the need to draw such distinctions.

2   Is programming a fine art or a useful art?

In Chapter 1, the structural similarities between software development and mathematics were introduced.[4] It may be recalled that according to the Curry-Howard isomorphism, the activities involved in doing mathematics map perfectly onto the activities involved in developing software. This being the case, the fact that mathematics is a fine art because it is aesthetic rather than rational, expressive rather than functional, and abstract rather than physical suggests that software deserves a similar classification. But the relationship between computer hardware and software is clearly a complicating factor. The question is thus whether the relationship thereby changes the nature of software development, or programming, such that it is properly considered a useful art.

To explore the patentability of software, the three-dimensional analytical framework applied to mathematics in the last chapter will be followed. That analysis requires a consideration of:

- whether software is an intangible or physical entity;
- whether software is an expressive or purposive artefact; and
- whether the creation of software is dominated by an aesthetic or rational approach.

Computer hardware, as the physical machine on which software executes, *prima facie*, has a direct impact on the physicality/intangibility dimension.[5] Therefore that dimension is considered first. However, it is conceivable that it may have an effect, directly or indirectly, on the other two dimensions of analysis. After consideration of each individual dimension, the nature of software development as a whole will be considered, before a concluded view is reached about the nature of software.

Before commencing that analysis, it is apposite to note that in this chapter, it is programming, rather than software, which is analysed. There are two reasons for

---

4 See page 30 and following.
5 That the execution of software on a general purpose computer creates a "new machine" tailored to a specific purpose "has some merit as a matter of computer science.": Pamela Samuelson, "Benson Revisited: The Case Against Patent Protection for Algorithms and Other Computer Program-Related Inventions," (1990) 39 *Emory Law Journal* 1025 at 1045, note 63. On the problems with this approach in Chapter 2, at 53. The problems with a purely physical analysis of computer software are also discussed in section 2.1 below.

this. The first emerges from the analysis of mathematics undertaken in Chapter 3. It may be recalled that there was an irreconcilable difficulty attending on a similar inquiry into the nature of mathematics. Various schools of philosophy had posited theories, none of which became dominant, and many of which were in direct conflict with each other. This lead to a decision to change the focus of the inquiry from what mathematics *is*, to what mathematicians *require*. It was through that change of focus that a useful distinction between patentable and non-patentable subject matter was derived in Chapter 4. Similarly, the focus in this chapter will not dwell on what software *is*, but on what programmers *require* to advance the art.

Looking at it another way, it is the activity of programming which is important, because it is that activity which is regulated by the grant of patent. Starting with a concept, and ending with executable code, the programmer determines the process or algorithm with precision, describing in detail the computable steps of the process.

2.1   Intangible or physical?

At the origins of modern computer, the "programmer's" task involved a physical reconfiguration of the hardware, plugging in wires and throwing switches. Such "programs" were "clearly physical and as much a part of the computer system as any other part."[6] Although the days of physical engagement with the hardware are gone, even in their modern form, computer programs can be understood physically, either as transforming a general purpose computer into a "new machine" running the specific program, manifested in the form of electrons pulsing through the circuitry of the computer.[7] Alternatively, they may be understood as a series of numbers stored in magnetic or other form on a carrier.[8] Indeed these characterisations are exactly those relied on by various courts, as discussed in Chapter 2.

However, computers generally, and computer software specifically, can be understood both at a physical level and at a symbolic level.[9] Indeed, "it is the understanding on the symbolic level which makes computers calculating devices, for it is under this kind of interpretation that various structures or processes of the computer are understood as symbols."[10] It is both pointless, and almost physically impossible, to attempt to describe the operation of a modern-day software package, such as Microsoft Word, by describing it by reference to the motion of electrons around the circuitry of the computer hardware. Even moving a level above to the realm of machine code, to attempt to describe Microsoft Word by

---

6 See James H. Moor, "Three Myths of Computer Science" (1978) 29(3) *British J Phil Sci* 213 at 213.

7 A discussion of this characterisation appears in Chapter 2, at page 53.

8 See Chapter 2, starting at page 57.

9 Moor, above n6 at 213, 215.

10 Moor, above n6 at 213.

detailing the logical operations performed by the CPU on bits stored in memory is futile. Even describing one particular function, such as counting the number of words in the currently open document, at this level is both difficult to comprehend, and failing to see the forest for the trees. Moving any further beyond this is to stray away from the purely physical into the realm of the symbolic.

Therefore a strictly physical analysis of software ignores the very essence of software. Brooks, in a seminal article addressing the failure of software to mark productivity advances in hardware,[11] captures the essence of software as "a construct of interlocking concepts; data sets, relationships among data items, algorithms and invocations of functions".[12] Generally speaking, these "objects of computer science are abstractions subject to logical – but not physical – constraints."[13]

A misunderstanding of this dual nature of software is also evident in case law in both the US and Australia which has used the term algorithm as if it is interchangeable with the term software.[14] Algorithms are better understood as either a marker for the purely symbolic aspects of a program;[15] the "mathematical counterpart of a textual object that is the program";[16] or as standing at a level of abstraction above programs, to the extent that programs are one of a class of possible implementations of a particular algorithm.[17] On any of these interpretations, the connection with the physical level is either absent, or significantly weakened.

The issue then is whether the symbolic or physical nature of software is of greater weight in the present context. The short history of software discussed in Chapter 1 demonstrates that the trend has been towards greater and greater abstractness, through

> the creation of tools, techniques, and computing hardware which permit programmers to be increasingly ignorant of the material realities of the machine, focusing instead on the abstractions they create and manipulate.[18]

---

11 Brooks notes that despite "desperate cries for a silver bullet – something to make software costs drop as rapidly as computer hardware costs do ... [t]here is no single development, in either technology or in management technique, that by itself promises even one order-of-magnitude improvement in productivity, in reliability, in simplicity.": Frederick P. Brooks Jr, "No Silver Bullet: Essence and Accidents of Software Engineering" (1987) 20(4) *Computer* 10 at 10.

12 Brooks, above n11 at 11.

13 Timothy Colburn and Gary Shute, "Abstraction, Law and Freedom in Computer Science" (2010) 41(3) *Metaphilosophy* 345 at 346.

14 The origins of this approach lie in *Gottschalk v Benson et al.* (1973) 409 U.S. 63.

15 Yiannis N. Moshovakis, "On founding the theory of algorithms", in Harold G. Dales and Gianluigi (eds), *Truth in Matemmatics* (Oxford University Press, 1998) 71 at 79.

16 Raymond Turner and Ammon Eden, "The Philosophy of Computer Science" *Stanford Encyclopedia of Philosophy* (Summer 2009 Edition) <http://plato.stanford.edu/archives/sum2009/entries/computer-science/> (25 July 2011).

17 See Moshovakis, above n15 at 75.

18 Scott Dexter et al., "On the Embodied Aesthetics of Code" (2011) 12 *Culture Machine* 1 at 11.

It might seem like sophistry to suggest that a process, software, executed on a physical device, a computer, and on which its execution can be observed, could nevertheless be characterised as non-physical. But this is not what is suggested. It is indeed possible to understand software at a purely physical level. But it must also be accepted that there is a relationship between the creation of software, and its execution.[19] But software cannot be understood only at that physical level. The more significant understanding of software for present purposes is on the symbolic level. To understand why, it is again useful to recall the difficulties attendant on determining the nature of mathematical objects.

But, as with mathematics, it is easy to become lost in an analysis of what software *is*. Given the isomorphism between software and mathematics, it is suggested that, just as with mathematics, what is *required* to advance the art is a much more useful avenue of inquiry.[20]

This being the case, to properly understand how a concern for the intangible emerges from the physical execution of software, and indeed comes to eclipse it, it is necessary to briefly diverge back into the history of mathematics. This is because it it was in the field of mathematics that the process by which mental processes might be externalised was developed. The emergence of the mechanical device by which such processes might be performed without human intervention therefore depends both conceptually and historically on these efforts.

A   *The externalisation of thought*

The ability to externalise or mechanise the mental processes owes its origins to the work of mathematicians such as Gottfried Leibniz and George Boole to "capture patterns of thought by means of algebra".[21] An (arguably) more modest goal was the later attempt by Bertrand Russell and Alfred North Whitehead to reduce all of mathematics to symbolic logic, in their work *Principia Mathematica*. That attempt was demonstrated to be a failure by Kurt Gödel. However, within his proof, Gödel used an ingenious device which deserves

---

19 The exact nature of that relationship has not been settled. See the discussion in Raymond Turner and Ammon Eden, "The Philosophy of Computer Science" *Stanford Encyclopedia of Philosophy* (Summer 2009 Edition) <http://plato.stanford.edu/archives/sum2009/entries/computer-science/> (25 July 2011) at section 2.1.

20 There is a correlation between the philosophy of mathematics and a philosophy of software here. To equate software with its physical instantiation is to take an empiricist view of software. A quasi-empiricist might however suggest that software is not *identical* with its physical instantiation, but is validated by it, with the symbolic aspects of software forming a web of belief around it. On the various internalist views, software *is* the abstract algorithms and data structures, which exist either only as products of the human mind, or externally in the Platonist realm. On these views, source code is merely a description of those objects, which as a side-benefit can be manifested in an executable form on a computer, if described precisely enough.

21 Keith Devlin, "The mathematics of human thought" in *Devlin's Angle*, Mathematical Association of America, January 2004 <http://www.maa.org/devlin/devlin_01_04.html> (22 July 2011).

further attention.[22] Gödel recognised that any formal language could be described as a string of unique symbols. For example,[23] the following statement

$$0 < S0$$

Contains the symbols "0", "<" and "S". Gödel's trick involves substituting for each well-formed statement a unique number.[24] Valid deductions within the mathematical system can then be "mapped" on to operations on those numbers, thereby reducing symbolic logic to mathematical calculation. It is this ability to understand something at both a symbolic level, and at the numerical level, which lies at the heart of computing. The computing hardware, operating as a mathematical calculation device, can then perform numerical calculations which are also capable of being understood at the symbolic level by humans.

The difference with computers and mathematics though, is that an additional layer of understanding exists below the numerical, wherein physical switches are represented as numbers, by assigning '1' to the on position and '0' to the off position.

When characterised in this way, it can be seen how the modern computer is an extension of the logicist programme. Hilbert's work on the *Eintscheidungsproblem* inspired both Gödel's work, his incompleteness theorem containing one of the first definitions of "general recursion", and also Turing's definition of "effective computability".[25] Although the links between Turing and Gödel are only circumstantial,[26] Turing's model of computation shows its historical origins in mathematics. This is nowhere more apparent than in Turing's comparison of "a man in the process of computing a ... number to a machine".[27] That

22 Gödel's proof is necessarily complicated. What follows is not a complete explication of the proof, but a summary of the techniques used which will be shown to be relevant to software development.

23 This example is taken from Wikipedia, "Proof sketch for Gödel's first incompleteness theorem" <http://en.wikipedia.org/wiki/Proof_sketch_for_G%C3%B6del's_first_incompleteness_theorem> (22 July 2011), which is in turn derived from the version used by Hofstadter in Douglas R Hofstadter, *Gödel, Escher, Bach: An Eternal Golden Braid* (Basic Books, 1980).

24 For example, if we assign to those symbols above, the numbers "666", "212" and "123", the statement above can be represented as 666 212 123 666.

25 See Martin Davis, *Computability and unsolvability* (Courier Dover Publications, 1982) at 11–12.

26 There is some evidence that Turing's work on computability was inspired by the work of both Gödel and Hilbert (see Edward R Griffor, *Handbook of computability theory* (Elsevier, 1999) at 11). There is also evidence of Gödel's direct influence on, but subsequent disapproval of, Church's work (see for example Griffor at 8–11). Finally, there is evidence that Gödel "enthusiastically accepted Turing's thesis and his analysis, and thereafter always gave credit to Turing (not to Church or to himself) for the definition of mechanical computability and computable function.": Griffor at 11.

27 Alan M. Turing, "On Computable Numbers, with an Application to the Entscheidungsproblem" (1936–1937) 42(2) *Proceedings of the London Mathematical Society* 230 at 231.

is, the original Turing machine was intended by Turing to model the process of *human* computation. Put another way:

> [B]y carrying out computations according to a selected plan, the mathematician acts in a way similar to a Turing machine: in considering some position in his writings and being in a certain "state of mind", he makes the necessary alterations in his writing, is inspired by a new "state of mind", and goes on to contemplate further writing. The fact that he completes more complicated steps than a Turing machine seems not principally significant.[28]

It will be recalled that this was exactly the basis on which the mental steps cases sought to exclude patentability, namely that a computerised process could also be carried out by a human operator. But the problem with the mental steps doctrine is that it tends to equate thought and computation:

> The ... mantra that cognition is computation, promulgated by early cognitive science researchers, defined thought in exactly those terms that could be instantiated by the digital computer, downplaying significant other aspects of the human mind, such as motivation, emotion and cross-cultural differences.[29]

In fact, Gödel's incompleteness theorem is a demonstration of the falsity of equating cognition and computation. His work demonstrated the difference between our intuitive understanding of concepts, and attempts to express them. This fuzziness is also evident in Turing's conception of what is computable. Thus it is asserted that the work of the programmer is in large part concerned with bridging this gap between the human conception of a process, and the precise description of a computable process in syntactically correct, executable machine form – the semantic gap.[30] Therefore, although it is true to say that computable processes can be understood at the physical level, and equivalently, that software can be understood as executing on computer hardware, the work of the programmer in most cases takes place a long way removed

---

28 SpringerLink, "Turing machines", in *Encyclopaedia of Mathematics* (2001) <http://eom. springer.de/t/t094460.htm> (22 July 2011).

29 Dexter et al., above n18 at 10. See also Peter Naur, "Computing versus human thinking" (2007) 50(1) *Communications of the ACM* 85; Philip J. Davis and Reuben Hersh *Descartes' Dream; The World According to Mathematics* (Dover Publications, 1986).

30 For an overview, see Wikipedia, "Semantic gap" <http://en.wikipedia.org/Semantic_gap> (25 July 2011). The semantic gap is particularly evident in natural language processing and image retrieval. On the latter see Arnold W.M. Smeulders et al., "Content-Based Image Retrieval at the end of the Early Years" (2000) 22(12) *IEEE Transactions on Patern Analysis and Machine Intelligence* 1 at 1: "There is something about Munch's 'The Scream' or Constable's 'Wivenoe Park' that no words can convey. It has to be seen. The same holds for a picture of the Kalahari desert, a dividing cell, or the facial expression of an actor playing King Lear. It is beyond words. ... Pictures have to be seen and searched as pictures: by object, by style, by purpose."

from these physicalities, and it is only in limited circumstances where software will be limited in any real sense by physicality.

The speed by which electronic computation is carried out makes possible more complex computations than would ever be practically reasonable for a human to carry out.[31] But this should not mean that the relationship between software and mental processes should be overlooked. Those who focus on the speed of the calculation itself overlook the sustained effort of the programmer to translate their own mental processes into a series of mechanical steps which can then be performed by a computer. Perlis captures that progression from mind to machine as follows:

> Every computer program is a model, hatched in the mind, of a real or mental process. These processes, arising from human experience and thought, are huge in number, intricate in detail, and at any time only partially understood. They are modelled to our permanent satisfaction rarely by our computer programs.[32]

The art of abstraction is the tool which is directed to closing this gap as much as possible. Just how much abstraction is appropriate involves a trade-off:

> You can 'ignore' the human concerns and end up with machine code ... You can ignore the machine and come up with a beautiful abstraction that can do anything at extraordinary cost and/or lack of intellectual rigor.[33]

Meaningful limitations: The key issue in the patentability context, is thus one of determining the extent to which a hardware device constrains the development of software in a meaningful sense. Thus the determination of whether claims over software are physical or intangible will involve difficult questions of degree. However, claims to a general purpose computer, or "any hardware" are unlikely, of themselves, to warrant a conclusion that some form of *meaningful* limitation is involved. It is only likely to be in limited circumstances where the physical aspects of a computer dictate are central to the design and implementation of the software which runs on it. It is suggested that size, performance or a need to communicate directly with specific hardware is when physical limitations

---

31 "Electronic computers are intended to carry out any definite rule of thumb process which could have been done by a human operator working in a disciplined but unintelligent manner. The electronic computer should however obtain its results very much more quickly.": Alan Turing, "Programmers' Handbook for Manchester Electronic Computer" *University of Manchester Computing Laboratory*, <http://www.AlanTuring.net/programmers_handbook> (21 July 2008).

32 Alan J. Perlis, "Foreword," in Harold Abelson and Gerald Jay Sussman, *Structure and Interpretation of Computer Programs* (2nd ed, MIT Press, 1996) at xi.

33 Bjarne Stroustrop in Federico Biancuzzi and Shane Warden, *Masterminds of Programming* (O'Reilly Media, 2009), at 5.

come into play. However Moore's law and practical experience suggest that it will be rare that the resource constraints exert any real limiting influence. Goodliffe suggests that the following contexts may require that the physical resources of the computer be a central concern:[34]

- game programming
- digital signal processing
- resource-constrained environments (for example deeply embedded systems)
- real-time systems[35]
- numerical programming.[36]

But there is a need to distinguish optimisation from physicality. The desirability of having code which efficiently executes on a computer is not necessarily the same as the physical limitations imposed on traditional inventions. This is because the physical limitation may be arbitrary, in the sense that optimisation is but one alternative,[37] and one which is best avoided, about 97% of the time.[38] Further, greater efficiency may be achieved by the use of a better design, or a better algorithm – that is, a non-physical solution can be employed.[39]

It is only really then when software must interact with a particular piece of hardware, and where the optimising compiler has no knowledge of the instruction set, that the instructions of the hardware device constitute a physical limitation on the design. The contexts in which this is likely to occur are in development of the lower levels of the operating system, such as kernel development; writing device drivers and bootloaders; or perhaps when writing platform-specific code for compilers or interpreters.[40]

B   *Summary*

In most cases, software is likely to be more intangible in nature than physical, although it may be sufficiently physical in a limited range of circumstances.

---

34 Pete Goodliffe, *Code craft: the practice of writing excellent code* (No Starch Press, 2007) at 205.
35 For example, in flight navigation systems or medical equipment.
36 For example some applications in the financial sector and scientific research require the processing of very large data sets.
37 Alternatives to optimisation might include: using faster hardware; optimising the various programs running on the hardware; asynchronously executing slow code; hiding the slow execution behind a responsive user interface; making the process run unattended; or using a new compiler: Goodliffe, above n34 at 204.
38 Donald E. Knuth, "Structured Programming with go to Statements" (1974) 6(4) *Computing Surveys* 261 at 268. This is because optimisation can impact on other desirable qualities of code, for example, readability, simplicity and maintainability or extensibility. See also Goodliffe, above n34 at 203.
39 The numerical programming examples listed above may well often fit into this category.
40 This will not always be the case though, as in many cases, some or all of a compiler might be written in a high level language. See the discussion of the Python language below on page 14.

2.2   Expressive or purposive?

Software is often written to achieve a particular purpose. It is common to define a computer program by reference to that function.[41] As noted above, software is big business, and software performs functions useful to all manner of enterprises, from controlling industrial processes to assisting academics to write large documents. Software products can be purchased as off the shelf components sold on the mass market, or written by a friend as a favour. The design, construction, and purchase of software would therefore seem to be dominated by a desire for function. Therefore it is tempting to place great weight on functionality. Indeed to the non-programmer, the notion of software as expressive may seem entirely at odds with their understanding of of it.

> In assessing any computer-related invention, it must be remembered that the programming is done in a computer language. The computer language is not a conjuration of some black art, it is simply a highly structured language. Analogously, if a person were to express a complete thought in German, it would be no trick for a translator to convert that thought into a palpable English form. The thought, thus expressed, might not be worthy of Shakespeare, but it would be understandable to one who uses the English language. Similarly, the conversion of a complete thought (as expressed in English and mathematics, i.e., the known input, the desired output, the mathematical expressions needed and the methods of using those expressions) into the language a machine understands is necessarily a mere clerical function to a skilled programmer.[42]

However, with a few examples it will be shown how software can express or communicate an idea. The programming language Perl is fertile ground for exploring this notion, since it is constructed in such a way that it is possible to write functioning code which directly engages visual perceptions. The high water mark of such code would have to be the program in Listing 5.1. This is, strictly speaking, a fully functional program, although its function may be of little economic value since it draws four smaller camel pictures, using its own visual layout as a template.[43] A similar example, set out in Listing 5.2,[44] demonstrates that it is possible to marry such visual expressiveness with useful functionality.

---

41 "A computer program is a set of instructions designed to cause a computer to perform a function or produce a particular result.": *Computer Edge v Apple* (1986) 161 CLR 177 at 178.

42 *In re Sherwood* 612 F.2d 809 at fn 6.

43 The code is also an interesting example of the difficulty of drawing distinct lines between code and data. If it is of no value, it might be wondered why the author went to the considerable difficulty of writing it. The value of writing such "toy programs" is considered in section E on page 197 below.

44 Taken from Alex Bowley <http://www.cs.cmu.edu/~dst/DeCSS/Gallery/bowley-efdtt-dvdlogo.html> (15 June 2008).

```perl
#!/usr/bin/perl -w                                    # camel code
use strict;

                                      $_='ev
                                   al("seek\040D
ATA,0,                             0;");foreach(1..3)
   {<DATA>;}my                     @camel1hump;my$camel;
 my$Camel  ;while(                 <DATA>){$_=sprintf("%-6
9s",$_);my@dromedary               1=split(//);if(defined($
_=<DATA>)){@camel1hum              p=split(//);}while(@dromeda
ry1){my$camel1hump=0               ;my$CAMEL=3;if(defined($_=shif
    t(@dromedary1                  ))&&/\S/){$camel1hump+=1<<$CAMEL;}
   $CAMEL--;if(d     efined($_=shift(@dromedary1))&&/\S/){
   $camel1hump+=1    <<$CAMEL;}$CAMEL--;if(defined($_=shift(
   @camel1hump))&&/\S/){$camel1hump+=1<<$CAMEL;}$CAMEL--;if(
   defined($_=shift(@camel1hump))&&/\S/){$camel1hump+=1<<$CAME
   L;;}$camel.=(split(//,"\040..m'{/J\047\134}L^7FX"))[$camel1h
   ump];}$camel.="\n";}@camel1hump=split(/\n/,$camel);foreach(@
   camel1hump){chomp;$Camel=$_;y/LJF7\173\175'\047/\061\062\063\
   064\065\066\067\070/;y/12345678/JL7F\175\173\047'/;$_=reverse;
   print"$_\040$Camel\n";}foreach(@camel1hump){chomp;$Camel=$_;y
   /LJF7\173\175'\047/12345678/;y/12345678/JL7F\175\173\0 47'/;
   $_=reverse;print"\040$_$Camel\n";}';;s/\s*//g;eval;     eval
     ("seek\040DATA,0,0;");undef$/;$_=<DATA>;s/\s*//g;(   );;s
       :^.* ;;;map{eval"print\"$ \""};/.{4}/g;   DATA   \124
     \1    50\145\040\165\163\145\040\157\1 46\040\1   41\0
             40\143\141   \155\145\1 54\040\1   51\155\   141
             \147\145\0   40\151\156  \040\141    \163\16 3\
           157\143\    151\141\16   4\151\1      57\156
           \040\167    \151\164\1    50\040\      120\1
           45\162\     154\040\15     1\163\      010\11
           1\040\1     64\162\1       41\144      \145\
           155\14      1\162\         153\04       0\157
            \146\       040\11        7\047        122\1
            45\15      1\154\1   54\171           \040
            \046\         012\101\16            3\16
            3\15         7\143\15             1\14
            1\16         4\145\163            \054
             \040          \111\156\14       3\056
             \040\        125\163\145\14      4\040\
             167\1       51\164\1   50\0     40\160\
           145\162                          \155\151
            \163\163                         \151\1
          57\156\056
```

Listing 5.1 Camel code

```c
#define m(i)(x[i]^s[i+84])<<

                unsigned char x[5]      ,y,s[2048];main(
                n){for( read(0,x,5      );read(0,s ,n=2048
                ); write(1    ,s,n)         )if(s
                [y=s      [13]%8+20]  /16%4   ==1      ){int
                i=m(      1)17   ^256 +m(0)    8,k       =m(2)
                0,j=      m(4)   17^ m(3)    9^k*      2-k%8
                ^8,a      =0,c    =26;for    (s[y]     -=16;
                --c;j    '-=2)a=     a*2^i&   1,i=i /2^j&1
                <<24;for(j=          127;      ++j<n;c=c>
                                     y)
                                     c

                +-y-i^i/8^i>>4^i>>12,
                i=i>>8^y<<17,a^=a>>14,y=a^a*8^a<<6,a=a
                >>8^y<<9,k=s[j],k        ="7Wo~'G_\216"[k
                &7]+2^"cr3sfw6v;*k+>/n."[k>>4]*2^k*257/
                8,s[j]=k^(k&k*2&34)*6^c+~y
                          ;}}
```

Listing 5.2 DeCSS code

This decryption program, known as DeCSS, deciphers an anti-copying measure embedded in the DVD format, known as region coding. The code above is a reformulation of the original DeCSS package, which was the subject of both criminal and civil litigation.[45] The algorithm has been expressed in a number of alternative forms, from mathematical descriptions,[46] to plain English,[47] to haiku,[48] and even as a prime number.[49] All of these reformulations, including the one in Listing 1.2, were written in response to the uncertainty around whether software, as a form of expression, should be entitled to constitutional protection under the First Amendment.[50]

The cynical reader may see these examples as nothing more than exceptions which make the rule. After all, the US Supreme Court has rightly noted that "it is possible to find some kernel of expression in almost any activity a person undertakes."[51] Indeed, there is some empirical evidence which suggests programmers see functionality as the dominant consideration.[52] On this basis it is tempting to suggest that software is "a fundamentally utilitarian construct even assuming it embodies some expressive element".[53]

Encryption software has been an important focal point for judicial consideration of the expressivity of software. And in that context, it is now accepted that at least one aspect of software, namely source code, is expressive. In *Bernstein v United States Department of Justice*,[54] it was held both at first instance, and on appeal, that source code was expressive and therefore protected by the right of freedom of expression contained in the First Amendment. Source

---

45 For a brief overview, see Ann Harrison, "DeCSS Creator Indicted in Norway" *Security Focus* 10 January 2002, <http://www.securityfocus.com/news/306> (27 June 2011). Movie studios in the US also successfully pursued distributors of the code, for breach of the DMCA provisions: see for example *Universal City Studios, Inc. v Corley* 273 F.3d 429 (2nd Cir. 2001).

46 See for example Charles M. Hannum, "A Mathematical Description of the CSS Cipher" <http://www.cs.cmu.edu/~dst/DeCSS/Gallery/hannum-pal.html> (15 July 2008).

47 See <http://www.cs.cmu.edu/~dst/DeCSS/Gallery/plain-english.html> (15 July 2008).

48 See <http://www.cs.cmu.edu/~dst/DeCSS/Gallery/decss-haiku.txt> (15 July 2008).

49 This 1401-digit number, when represented in binary form is a zipped copy of the original C source code of the descrambler. See <http://www.utm.edu/research/primes/curios/48565...29443.html> (15 July 2008).

50 The DeCSS campaign was sparked by a preliminary ruling by District Court Judge Kaplan in *MPAA v Reimerdes, Corley and Kazan* (NY, 2 February 2000) which suggested firstly that it was "far from clear" that source code was expressive, and that even assuming some expressivity in source code, "the expressive aspect appears to be minimal when compared to its functional component". Cf. John D. Touretzky, "Gallery of DeCSS Descramblers" (2000) <http://www.cs.cmu.edu/~dst/DeCSS/Gallery/> (11 July 2011).

51 *City of Dallas v Stanglin* 490 US 19 (1989) at 25.

52 Dexter et al., above n18 at 6, citing A. Kozbelt et al. "Beautiful Software: Characterizing Aesthetic Judgment Criteria of Code Amongst Expert and Novice Computer Programmers", paper presented to 2010 Biannual Meeting of the International Association of Empirical Aesthetics, Dresden, Germany.

53 *Universal City Studios Inc v Reimerdes* 82 F. Supp.2d 211 (SDNY, 2000) at 226 per Judge Kaplan.

54 176 F.3d 1132 (9th Circuit, 1999) (*"Bernstein"*).

code was "protected speech" because "[t]he distinguishing feature of source code is that it is meant to be read and understood by humans and that it can be used to express an idea or a method."[55] Indeed, some programmers see this communicative aspect as the most important consideration in programming.[56]

The use of source code as a way of communicating and representing ideas was also recognised in *Junger v Daley*:

> The issue of whether or not the First Amendment protects encryption source code is a difficult one because source code has both an expressive feature and a functional feature. The United States does not dispute that it is possible to use encryption source code to represent and convey information and ideas about cryptography and that encryption source code can be used by programmers and scholars for such informational purposes. Much like a mathematical or scientific formula, one can describe the function and design of encryption software by a prose explanation; however, for individuals fluent in a computer programming language, source code is the most efficient and precise means by which to communicate ideas about cryptography.
>
> ...
>
> Because computer source code is an expressive means for the exchange of information and ideas about computer programming, we hold that it is protected by the First Amendment.[57]

Eventually source code is translated to a form which is executable on a particular piece of computer hardware. But it maintains its descriptiveness of that computable process. This question then arises: by the time it has taken its executable form, has software has lost its expressiveness? This issue has been dealt with in the copyright context. A suitable example from the Australian jurisdiction is *Computer Edge v Apple*.[58] At trial it was held that object code was not suitable subject matter for copyright, because it did not fit the definition of a literary work, not being "intended to afford information, instruction or pleasure in the form of literary enjoyment".[59]

On appeal to the Full Federal Court, source code was held to "express meaning as to the arrangement and ordering of instructions for the storage

---

55 *Bernstein* at 1140.

56 "Programs must be written for people to read, and only incidentally for machines to execute.": Harold Abelson and Gerald Jay Sussman, *Structure and Interpretation of Computer Programs* (2nd ed, MIT Press, 1996) at xvii.

57 209 F.3d 481 (2000) at 485.

58 *Apple Computer Inc v Computer Edge Pty Ltd* (1983) 1 IPR 353, 50 ALR 581 ("*Computer Edge (FCA)*") (first instance); *Apple Computer Inc v Computer Edge Pty Ltd* (1984) 1 FCR 549 (Full Federal Court) ("*Computer Edge (FCAFC)*"); *Computer Edge Pty Ltd v Apple Computer Inc* (1986) 161 CLR 171 (High Court) ("*Computer Edge* (HC)").

59 *Computer Edge (HC)* citing in support *Hollinrake v Truswell* [1894] 3 Ch. 420; *Exxon Corporation v Exxon Insurance Consultants International Ltd* [1982] R.P.C. 69 at 88.

and reproduction of knowledge. It is incorrect to describe them simply as components of a machine."[60] The Full Court held that

> [t]here is no necessity for a literary work to be of any literary quality. It is accepted that the term includes mathematical tables, codes, and, in general, alphanumerical works. One limit doubtless is that it needs to be a "work" and to have had some skill, even if very small, applied to its preparation. Meaningless rubbish would plainly be excluded.[61]

The Full Court took the view that it did not matter whether object code was a literary work or not, "because [it] can fairly be described as [a] translation"[62] On further appeal, the High Court reinforced the position, holding that source code was akin to a literary work, but object code was not:

> [The object] programs existed in the form of a sequence of electrical impulses, or possibly in the pattern of circuits that when activated generated those electrical impulses. On any view they were not expressed in writing or print. Although the electrical impulses could be represented by words or figures, the impulses themselves did not represent or reproduce any words and figures. They were not visible or otherwise perceptible, and they were not, and were not intended to be, capable by themsleves of conveying a meaning which could be understood by human beings. ... It is true that the object programs might have been printed out in binary or hexadecimal form, but the question whether any such written expression of the programs would have been a literary work is not the question that now falls for decision. We are concerned with the object programs embodied in the ROMS and it seems clearly to follow from the cases already cited, which decide that a literary work is a work expressed in print or writing, that they were not literary works.[63]

Thus it might be argued that the tendency is to emphasise the expressiveness as located in source code, and not present in object code, which is functional. The solution in Australia has been to amend the *Copyright Act 1968* to expressly include object code within the definition of a literary work.

The US position, which proceeds from a similar theoretical position,[64] provides an interesting counterpoint. For example, in *Williams Electronics v Artic International*[65] the defendant to a claim for infringement asserted just such a

60 *Computer Edge (FCAFC)* at 558–559 per Fox J.
61 *Computer Edge (FCAFC)* at 558 per Fox J.
62 *Computer Edge (FCAFC)* at 559 per Fox J.
63 *Computer Edge (HC)* at 183 per Gibbs CJ.
64 So too the UK position. See *SAS Institute Inc v World Programming Ltd* [2010] EWHC 1829 (Ch) in which it was held that the WIPO treaty, Berne Convention and TRIPS all support the distinction between copyrightable expression and unprotectable ideas.
65 *Williams Electronics v Artic International* 685 F.2d 870 (3rd Circuit, 1986) ("*Williams*").

distinction between expressive source code and functional object code. However, the court declined to draw such a distinction, holding that object code came within the terms of the US Copyright Act.[66] This is in no small part due to the legislative definition by which a "copy"

> include[s] a material object in which a work is fixed "by any method now known or later developed, and from which the work can be perceived, repro- duced, or otherwise communicated, either directly or with the aid of a machine or device." By this broad language, Congress opted for an expansive interpretation of the terms "fixation" and "copy" which encompass techno- logical advances such as those represented by the electronic devices in this case.[67]

The Court in *Apple v Franklin*[68] approached the same issue from a slightly dif- ferent angle, invoking the *Baker v Selden*[69] case as authority for distinguishing between copyright protection for the "mere" instructions as a literary work on the one hand, and protecting the method behind the instructions as a utilitarian, or functional, work. The issue in patent law is perhaps reversed, since it is asserted that expressiveness is an indicium of a non-patentable fine art. To that end, it is worth exploring further the extent to which all code, even the executable code most aligned with function, achieves a level of expressiveness.

A starting point in assessing the expressiveness of code is to say something more about the medium in which software is constructed – the programming *language*. A non-programmer might be tempted by the the view that such "lan- guages" are no more than convenient standards by which to interface to, or direct, the computer's machine-code operations.[70] Put another way, a language might be seen as nothing more than a "fixed collection of rules"[71] to be assem- bled in a mechanical manner to achieve the required function. If these languages were mere "standards" we might expect that there were only a few of them, in much the same way that operating systems have "standardised" around two major variations – Windows and Unix.[72]

But a computer language is much more than an objective standard. It is "a system of notation for describing computation"[73] with an underlying structure

---

66 *Copyright Act* 17 USC (1976).
67 *Williams* at 876–877.
68 714 F.2d 1240 (3rd Circuit, 1983).
69 101 U.S. 99 (1879).
70 Paul Graham "Great Hackers" <http://www.paulgraham.com/gh.html> (16 June 2011).
71 Hal Abelson, "Foreword" in Daniel P. Friedman, Mitchell Wand and Christopher T. Haynes, *Essentials of Programming Languages* (MIT Press, 2001) at vii.
72 This reductionist view might itself be challenged. First, Unix includes a large number of var- iants, including a number of GNU/Linux and BSD distributions, and MacOSX. Second, although Unix variants share a design, Windows variants only share a trademark, and a certain degree of backward compatibility with earlier brand members.
73 Robert D. Tennent, *Principles of Programming Languages* (Prentice Hall, 1990) at 1. See also *Funger v Daley*.

that reflects the language designer's decisions.[74] These decisions are a compromise between conflicting goals:

> A useful programming language must ... be suited for both description (i.e., for human writers and readers of programs) and for computation (i.e., for efficient implementation on computers). But human beings and computers are so different that it is difficult to find notational devices that are well suited to the capabilities of both.[75]

The scope for variation between the formal specification of a particular language, (discussed below in relation to Python and C#) also suggests the appropriateness of analogies with other expressive works, such as literary and other copyrightable works. Just as there are multiple ways to write a book given a specific storyline,[76] there are myriad ways to give effect to the function of an algorithm, and programmers at least see value in exploring different expressions. If a language is merely a standard, surely there would be no need for more than one implementation of it. This is suggestive of the value which lies in the expression of computable processes in a variety of ways.[77]

Also influencing language design is the fact that programmers have their own styles that reflect the way they think,[78] and the capacity of languages to influence the way programmers think makes language choice an important decision.[79]

---

74 Abelson, above n71 at vii.
75 Tennent, above n73 at 1.
76 In fact, it is suggested at various times that there are only a very small number of story lines to choose from, from 1 to 30. See IPL2, "The 'Basic' Plots in Literature" <http://www.ipl. org/div/farq/plotFARQ.html> (8 July 2011).
77 The notion of value in programming is explored below in Section 2.3 on page 185.
78 "[Language] syntax deeply affects how you think about the problem, even though semantically it has absolutely no bearing on what's going on." Anders Hejlsberg, designer of the C# language, in John Osborn, "C#: Yesterday, Today, and Tomorrow: An Interview with Anders Hejlsberg, Part 2" *OnDotNet*, 31 October 2005 <http://ondotnet.com/pub/a/ dotnet/2005/10/31/interview-with-anders-hejlsberg-part-2.html>(8 July 2011).
79 Dijkstra was an ardent proponent of the influence of computer languages on thought:

> The tools we use have a profound (and devious!) influence on our thinking habits, and, therefore, on our thinking abilities.
>
>   ...
>
>   It is practically impossible to teach good programming to students that have had a prior exposure to BASIC: as potential programmers they are mentally mutilated beyond hope of regeneration.
>   The use of COBOL cripples the mind; its teaching should, therefore, be regarded as a criminal offence."

> Edsger W. Dijkstra, "How do we tell truths that might hurt?", (EWD498, 1975) <http:// www.cs.utexas.edu/~EWD/transcriptions/EWD04xx/EWD498.html> (29 July 2011), reproduced in *Selected Writings on Computing: A Personal Perspective* (Springer-Verlag) at 129–131. See also Robin Milner, designer of the ML language, cited in Dexter et al., above n18 at 12.

For example, the Perl language is designed around the idea that "there's more than one way to do it."[80] In contrast, the Python language's philosophy is that "there should be one – and preferably only one – obvious way to do things."[81]

What these influences on language design also demonstrate, is that a programming language may be thought of as just another program. In fact, Abelson describes this very notion as "the most fundamental idea in computer programming":[82]

> Consider … the basic idea: the interpreter itself is just a program. But that program is written in some language, whose interpreter is itself just a program written in some language whose interpreter is itself.[83]

This recursiveness can be demonstrated by looking at the Python language. On a purely abstract level, Python can be thought of as a mere collection rules of syntax and semantics.[84] Giving effect to that abstract specification is are a number of implementations. The "main" implementation of Python is written in C, and is sometimes referred to as CPython. There are also implementations written in Java (Jython)[85] and C# (IronPython).[86] There is even an implementation written in Python itself (PyPy).[87] Digging a level deeper, the IronPython interpreter is written in C#, a language developed by Microsoft, which is written in C++,[88] and compiles to an intermediate language called MSIL.[89] MSIL can be compiled to machine code beforehand, or executed by an machine-code interpreter at runtime, these compilers/interpreters are "likely to have been written

---

80 Wikipedia, "There's more than one way to do it" <http://en.wikipedia.org/wiki/There's_more_than_one_way_to_do_it> (11 July 2011).

81 Tim Peters, "PEP 20 – The Zen of Python" <http://www.python.org/dev/peps/pep-0020/> (7 July 2011). The inventor of the language, and lead programmer, Guido van Rossum, when asked about the origin of this design goal, noted that it "comes straight from the general desire for elegance in mathematics and computer science", being an example of orthogonality: Frederico Biancuzzi and Shane Warden, *Masterminds of Programming* (O'Reilly Media, 2009) at 25.

82 Abelson, above n71 at vii.

83 Abelson, above n71 at viii.

84 Although there is no rigorous formal specification of every aspect of the language, the Python Language Reference is considered "the" guide. See "The Python Language Reference" <http://docs.python.org/reference/index.html> (11 June 2011).

85 See <http://www.jython.org/> (11 June 2011).

86 <http://ironpython.net> (8 July 2011).

87 <http://pypy.org/> (11 June 2011). Implementing a language in itself is known as bootstrapping, or self-hosting: Wikipedia, "Bootstrapping (compilers)" <http://en.wikipedia.org/wiki/Bootstrapping_(compilers)> (27 May 2011).

88 At the very least, the shared source implementation of C# is. See "Shared Source Common Language Infrastructure 2.0 Release" <http://www.microsoft.com/download/en/details.aspx?displaylang=en&id=4917> (8 July 2011).

89 This stands for Microsoft Intermediate Language. See Microsoft, "Managed Execution Framework" <http://msdn.microsoft.com/en-us/library/k5532s8a.aspx> (8 July 2011).

in Microsoft Visual C."[90] To actually run, the executable object code depends on a framework called the Common Language Framework, which was written in C++, although it seems that at least one component was prototyped in Lisp.[91] The Microsoft specifications for C#, MSIL and machine code interpreters have been independently implemented by an open source project called Mono, which is entirely implemented in C#.[92] Mono is available for Mac OSX, Windows and various Linux distributions.[93]

Putting aside the distraction of all the technical details, this short overview serves to demonstrate both how a language is a program, assembled from a series of component parts, in much the same way as any modern application can be constructed from the building blocks of a language, various libraries, and frameworks.[94]

The example also demonstrates how the division between source code and object code is not cleanly drawn. Ultimately, the form into which the source code is translated *is* one in which it is capable of actually causing the computer to carry out the described process. But that path to a final form "a complex and remarkable chain of events",[95] and may not be completed until just before the time the code is executed. Any claims to a clean separation between expressive source code and functional object code must therefore be cautiously approached.

It also demonstrates how code can be seen to be *executable data* – the source code for software, rather than being seen as a mere adjunct to the process as executing, can be understood as the precise description of a computable process. That description forms the input (data) to a program which translates it to another form. The output of that program may be into machine code, or it may be to another intermediate language requiring further translation before execution. The software also may not reach its final, executable state until just before it is in fact executed.

An important corollary of the interpreter as a program should also be considered. Any program can also be conceived of as a language, a collection of languages, or at the very least as containing "important [language]-like pieces."[96]

90 Because this is a proprietary program, the source code is not available. However, the likelihood of the compiler, ngen.exe having been written in C is asserted in FaultWire's analysis. See FaultWire, "Ngen.exe CLR Native Compiler" <https://www.faultwire.com/file_detail/ngen.exe*41826.html> (8 July 2011).

91 "I designed the architecture of the runtime and wrote the Garbage Collector (and yes the GC prototype was written in Common Lisp first and I wrote a translator to convert it to C++)": Patrick Dussud, "How It All Started…AKA the Birth of the CLR" on *CLR, Architectures and Stuff* <http://blogs.msdn.com/b/patrick_dussud/archive/2006/11/21/how-it-all-started-aka-the-birth-of-the-clr.aspx> (8 July 2011).

92 Mono Project, "CSharp Compiler" <http://www.mono-project.com/CSharp_Compiler> (8 July 2011).

93 See <http://www.mono-project.com/download/> (21 August 2017).

94 See the discussion in Chapter 1 at 18.

95 Colburn and Shute, above n13 at 355.

96 Abelson, above n71 at viii.

Larry Wall, designer of the Perl programming language, adverts to this when he suggests that "if you think of Unix as a programming language, it's far richer than even Perl. Perl is, by and large, a digested and simplified version of Unix. Perl is the Cliff Notes of Unix."[97] Abelson advocates exactly such an understanding, suggesting that

> [o]ne of the most powerful ways to structure a complex program is as a collection of languages, each of which provides a different perspective, a different way of working with the program elements.[98]

In the absence of a clear division between source code and object code, Colburn and Shute suggest a way of characterising the expressiveness of software as a function of the level of abstraction involved:

> As levels of programming language abstraction increase, the languages become more expressive in the sense that programmers can manipulate direct analogs of objects that populate the world they are modeling, like shopping carts, chat rooms, and basketball teams. This expressiveness is only possible by hiding the complexity of the interaction patterns occurring at lower levels.[99]

This suggests that expressiveness is determined by physicality, being proportional to the extent to which physical restrictions impose themselves on the programming process. Such a view certainly lends further weight to the analysis set out in relation to physicality above, and in particular the need for careful characterisation of the invention.

In this context it may also be useful to determine the baseline: how expressive is the most functional of all code, that is, code in its final machine-executable form? In Chapter 2 it was asserted that the "new machine" characterisation of software, whilst correct as a matter of computer science, is nevertheless problematic.[100] One reason for this is that such a characterisation at best overlooks, and at worst strips away, an essential characteristic of software:

> Because the operations of these machines have no apparent meaning outside of human activity, it is easy to conclude … that the meaning of a program is identical to the human-interpreted result of its function: once the 'internal' mental symbols are brought outside the body, the crucial connection with 'external' states is severed, the 'internal' mental symbols all but vanish as they are stripped of their meaning, and we are left with machines which,

---

97 Larry Wall, "Perl, the first postmodern computer language" <http://www.perl.com/pub/1999/03/pm.html> (7 July 2011).
98 Abelson, above n71 at viii.
99 Colburn and Shute, above n13 at 348.
100 See Chapter 2, at 53.

with a compelling illusion of near-autonomy, traverse a wide range of meaning-laden states.[101]

Put another way, software is not adequately described by its mere functionality, "existing outside of and independent from human experience"[102] but is inextricably linked to its human meaning. There is a temptation to dismiss such an argument on the basis that any technical or engineering process is imbued with just such a kernel of expression. Expressivity can operate to positively enhance the perception of functionality. A Mac fanboi[103] purchases a MacBook not only because of its technical specifications, like processor speed and memory capacity, but because of what it *means* to be a Mac owner. It might be a statement about valuing Apple's "it just works" philosophy,[104] a commitment to industrial design, or even a protest against the dominance of Windows in the operating systems market.

This grounding of meaning in what might at first blush be thought to be predominantly functional opens the door to an understanding of the ways in which software is "technically expressive". Ratto demonstrates how software may contain of "embedded technological expression"[105] in three ways. First, as touched on above, the source code of a program expresses, or explains, the process whereby particular tasks can be accomplished.[106] Second, software when run on a computer expresses to users how they may interact with the software.[107] Third, the modularisation of a program undertaken as a part of the design process acts as a method of organising labour. That is, it defines "relationships between programmers".[108]

---

101 Dexter et al., above n18 at 10.
102 Dexter et al., above n18 at 9.
103 A fanboi is a pejorative term for "a person with an irrational attachment to a particular item or brand name, and an equally irrational dislike for competing brands or items.": "fanboi or fanbois", *Whirlpool Knowledge Base* <http://whirlpool.net.au/wiki/fanboi> (4 August 2011); "fanboi" *Urban Dictionary* <http://www.urbandictionary.com/define.php?term=fanboi> (4 August 2011).
104 See for example Chad Dickerson, "Mac OSX: it just works" *Infoworld*, 12 September 2003 <http://www.infoworld.com/t/platforms/mac-os-x-it-just-works-441> (4 August 2011). The phrase was originally an Apple slogan for its Unix-based operating system, OSX. See Wikipedia, "List of Apple Inc. slogans" <http://en.wikipedia.org/wiki/List_of_Apple_Inc._slogans> (4 August 2011).
105 Dan L. Burk, "Patenting speech," (2000) 79 *Texas Law Review* 99 at 115; Matt Ratto, "Leveraging Software, Advocating Ideology: Free Software and Open Source" paper presented to the 29th Research Conference on Information, Communication and Internet Policy (2001) <http://arxiv.org/abs/cs.CY/0109103> (27 June 2011) at 12.
106 Matt Ratto, "Leveraging Software, Advocating Ideology: Free Software and Open Source" paper presented to the 29th Research Conference on Information, Communication and Internet Policy (2001) <http://arxiv.org/abs/cs.CY/0109103> (27 June 2011) at 12.
107 Ratto, above n106 at 14.
108 Ratto, above n106. The relationship between organisational structure and the structure of code is also reflected in Conway's Law, that "[a]ny organization that designs a system will produce a design whose structure is a copy of the organization's communication

Dijkstra believed that expression is central to the activity of programming, although in a way which extends both beyond the mere expression of a particular idea in a programming language, and the notions of technical expression set out above. Dijkstra believed the need for natural language mastery in programming was twofold: first, it was necessary for the accurate communication and description of the problem to be solved, and also as an "indispensible tool for thinking".[109]

A   Summary

The classical approach to determining the appropriate protection for software is argued along this dimension. One the one hand, copyright is said to protect expressive works, and patent law to protect functional works. However programs are "machines whose medium of construction is text"[110] and exhibit a duality, or deep interdependence, of functionality and expressiveness which makes them hard to classify.

Attempts to shoehorn software into one or other category have therefore faced difficulties. In the copyright context, the expressive nature of source code is emphasised as making it akin to a literary work. However, in recognition of the fact that there is a tight causal relationship between source code and object code, copyright regimes have been expanded to cover object code, which shares little similarity with other copyrightable works, in that it *behaves*.

Similarly, in the patent law context, a focus on the function of software has a similarly damaging effect. There is no doubt that software is functional. To assert otherwise would be folly. However, software is expressive, in at least the three ways just described. So if anything, the choice in this context might be between software as functionally expressive or expressively functional. Given the way in which it has been demonstrated that software's function is also expressive, the latter of these terms would seem the most appropriate.

The result of allowing patenting of subject matter which strays into the realm of the expressive, has meant the traditional limitations on patent law have been watered down to the point where the regime has allowed patents over movie scripts, sports moves, and tax advice. This fact may of itself suggest the appropriateness of finding a solution to the software patent problem which is reflective of its unique character – a *sui generis* regime.

2.3   Aesthetic or rational?

Looking at the design process for software, and recalling the differences attributed to artistic and engineering design by Mitcham by reference to their "ends

---

structure.": <http://beautifulcode.oreillynet.com/2007/11/hidden_influences_on_software.php> (15 July 2008).

109 E. W. Dijkstra "Programming as a Discipline of Mathematical Nature" (1974) 81(6) *The American Mathematical Monthly* 608–612.

110 Pamela Samuelson et al., "A Manifesto Concerning the Legal Protection of Computer Programs," (1994) 94 *Columbia Law Review* 2308 at 2320.

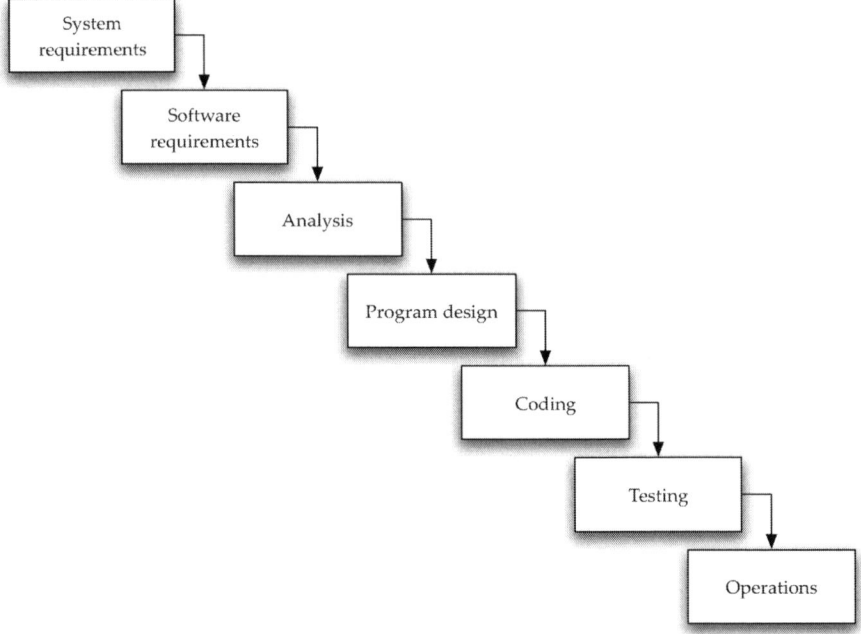

Figure 5.1 The idealised waterfall development model

in view",[111] it is possible to determine whether software is primarily directed to efficiency or beauty. That is, whether it is a rational or aesthetic creation. This may be a question of degree to some extent, since it is clear that even in traditional engineering disciplines such as building of bridges and towers, "structural engineers are guided by aesthetics as well as [rational] analysis."[112] However, whilst both aesthetic and rational considerations might be at play in software development, it is submitted that the aesthetic aspect is so integral to software development that it cannot be considered a mere ancillary factor, as it might be in other technological endeavours.

A rational engineering approach involves a sequence of logical steps. A good example of this model in action is the idealised waterfall model of software development, set out in Figure 5.1. This approach to software development dominated in the early years of software development, and was a product of the dominance of hardware engineering, and hence a hardware engineering

---

111 See Chapter 4, at 40.
112 Bruce J. MacLennan, "Aesthetics in Software Engineering" (Technical Report No UT-CS-06-579, Department of Electrical Engineering and Computer Science, University of Tennessee, 31 October 2006) <http://web.eecs.utk.edu/~library/TechReports/2006/ut-cs-06-579.pdf> (26 May 2011).

approach.[113] Under the waterfall model, each phase of the design and development process ought be completed before the next is begun, with some small tolerance for interaction between levels.

But this ideal of a rational design process has never matched the reality of software development, and never will. At best, it will be seen that this model is something which might be worth "faking",[114] in the same way that the rigorous logical deduction from axiom to proof in a published mathematical proof belies the tortuous and winding path of discovery which precedes it.[115] This "tortuous path" is the hallmark of the irrational, emotional, intuitive, human, or aesthetic mode of creation. It is how books are written, how a block of stone becomes a sculpture, how an artist moves from an idea to a work in a gallery.

A   *Just another type of engineering?*

The development of software might at first blush seem primarily rational. In many cases "the [software development] problem is appropriately an engineering one: creating cost-effective solutions to practical problems, building things in the service of mankind."[116] It is directed towards the assembly of a sequence of logical steps, syntactically and grammatically correct according to the strictures of the "language" in which it is written. Not only must it in fact be correct in order to compile and run, thereby transforming the computer into a "new machine",[117] but the eventual "product" will be expected to efficiently utilise the available resources of the computer to perform the functions which it was written to perform. As noted by Mitcham, this notion of efficiency is at the heart of engineering design.[118]

Also, many programmers are likely to see themselves as engineers, working through a series of rational steps towards a defined end goal. This link with

113 Barry Boehm, "A View of 20th and 21st Century Software Engineering" (Paper presented at the 28th International Conference on Software Engineering, Shanghai, China, 20 May 2006) 12 <http://portal.acm.org/citation.cfm?id=1134288> at 13.

114 David L. Parnas and Paul C. Clements, "A Rational Design Process: How and Why To Fake It" (1986) 12(2) IEEE Transactions on Software Engineering 251.

115 Parnas and Clements, above n114 at 251; Daniel M Berry, "The Inevitable Pain of Software Development: Why There Is No Silver Bullet" in Martin Wirsing, Alexander Knapp and Simonetta Balsamo (eds), *Radical Innovations of Software and Systems Engineering in the Future, Proceedings of the 2002 Monterey Conference*, LNCS 2941 (Springer, 2004) 50 at 53.

116 Mary Shaw, "Prospects for an Engineering Discipline of Software" (1990) 7(6) *IEEE Software* 15, at 19.

117 For a discussion of the new machine characterisation of software, see Chapter 4, at 53.

118 Cf. H. Coqui, "Corporate survival: The software dimension" Focus '89, Cannes, 1989, cited in Frederick P. Brooks Jr, *The Mythical Man Month: Essays on Software Engineering* (Anniversary ed, Addison-Wesley, 2002) at 218: "The driving force to use Software Engineering principles in software production was the fear of major accidents that might be caused by having uncontrollable artists responsible for the development of ever more complex systems."

engineering has a historical basis. When software first emerged in the 1950s it was, understandably, closely aligned with computer hardware engineering.[119] In the 1960s however, the demand for software coupled with the emerging "hacker" culture saw software development break free of this close association with hardware engineering, into a discipline in its own right.[120] At the same time, the growing complexity of software projects, the increasing ambition of programmers, and a series of colossal failures on software development projects prompted NASA to hold two conferences on the emerging "software crisis".[121] In response to this crisis, the term "software engineering" was coined in 1968.[122] Software engineering, like any other engineering discipline, would need to proceed on a disciplined, rational basis, in order to achieve the productivity gains comparable to those being achieved by computer hardware engineers.[123]

This view of software development as a rational process might also be reinforced by the fact that software development is an industry of enormous economic importance. Software products keep getting bigger, more complex, and require better organisation and management in order to be developed. Organisations developing software solutions therefore stand to benefit from reinforcing the stable, "engineering" nature of software development in order to assure customers and investors that they are making a safe bet.[124] However, it is submitted that such project management and business process issues lie beyond any properly defined software engineering scope.[125]

B   No silver bullet

It was noted that the historical schism between programming and hardware engineering emerged in the 1960s. Despite the attempts to recharacterise software development as an engineering discipline, in 2009, Shaw acknowledged that no practice of software engineering had yet emerged.[126] Whilst new

---

119 Boehm, above n113 at 13.
120 Boehm, above n113 at 13–14.
121 Boehm, above n113 at 14. Dijkstra asserted that the crisis was directly referable to the nature of the programming activity: Edsger W. Dijkstra, "The Humble Programmer", (1972) 15 (10) *Communications of the ACM* 859 at 862.
122 Edsger W. Dijkstra, "The Humble Programmer", (1972) 15 (10) *Communications of the ACM* 859 at 14.
123 Brooks, above n11 at 10.
124 Andrei P. Ershov, "Aesthetics and the Human Factor in Programming" (1972) 15(7) *Communications of the ACM* 501.
125 "[W]henever someone says 'we don't need technical advances; we need a better process," that's a sign that production skills haven't yet brought us to a fully mature commercial practice.": Mary Shaw, "Continuing Prospects for an Engineering Discipline of Software" (2009) 26(6) *IEEE Software* 64 at 66. According to Shaw, a mature commercial practice is a pre-cursor to the emergence of an engineering practice.
126 Shaw, above n125.

methodologies are regularly said to herald the dawn of the software engineering age,[127] these so-called silver bullets[128] always miss their mark.

Further, there are good reasons to think that a true engineering practice in software development may never take hold. The main reason for adopting such a negative view stems from the inherent nature of the software development task. Engineering practice limits the impact of variance between individuals, by enabling "ordinary practitioners so they can create sophisticated systems that work – unspectacularly perhaps, but reliably".[129] Where an activity is objective, or rational, this is easily done. But "[t]he essence of a software entity is a construct of interlocking concepts; data sets, relationships among data items, algorithms and invocations of functions."[130] As such, "the hard part of building software [is] the specification, design and testing of this conceptual construct, not the labor of representing it and testing the fidelity of the representation."[131] This distinguishes software practice from engineering practice:

> Whereas the difference between poor conceptual designs and good ones may lie in the soundness of design method, the difference between good designs and great ones surely does not.[132]

Correspondingly, there is some empirical evidence to suggest that programmers of similar experience may exhibit differences in productivity of an order of magnitude.[133] As an endeavour dependent on individual attributes, it is submitted that aesthetic considerations must play a central role.

---

127 See Michael A. Dryja, "Looking to the Changing Nature of Software Development for Clues to Its Protection" (1995) 3 *University of Baltimore Intellectual Property Law Journal* 109. Shaw takes a positive, yet more guarded position, arguing that there "are good grounds to to expect that there will eventually be an engineering discipline of software.": Shaw, above n116 at 22.

128 "Of all the monsters that fill the nightmares of our folklore, none terrify more than werewolves, because they transform unexpectedly from the familiar into horrors. For these, one seeks bullets of silver that can magically lay them to rest.

The familiar software project, at least as seen by the nontechnical manager, has something of this character; it is usually innocent and straightforward, but is capable of becoming a monster of missed schedules, blown budgets, and flawed products. So we hear desparate cries for a silver bullet – something to make software costs drop as rapidly as computer hardware costs do.": Brooks, above n11 at 10.

129 Shaw, above n116 at 16.

130 Brooks, above n11 at 11.

131 Brooks, above n11 at 11.

132 Brooks, above n11 at 18.

133 H. Sackman, W. J. Erickson and E. E. Grant, "Exploratory Experimental Studies Comparing Online and Offline Programming Performance" (1968) 11(1) *Communications of the ACM* 3; Bill Curtis, "Fifteen Years of Psychology in Software Engineering: Individual Differences and Cognitive Science" ICSE '84 Proceedings of the 7th international conference on Software engineering (IEEE Press, 1984) <http://portal.acm.org/citation.cfm?id=801956> (27 July 2011) 97 at 98–99; Sarah H. Nash and Samuel T. Redwine Jr,

Beynon et al. note that "[o]ne of the main purposes of adopting a formal approach to program development is to shift the focus from 'considering your intuitive ideas about what a program is intended to do' to 'thinking precisely and abstractly about what is it that you have instructed it to do'."[134] Viewed as such, emphasising such a structured approach to software development address the potential hazard which exists "in the relation between what we can see immediately, and perhaps superficially, and what we might be able to see upon closer scrutiny in other contexts, possibly unforseen."[135]

However, a formal approach, whilst perhaps an important teaching tool, does not remove intuition, which may still operate in the interpretation of a formal specification.[136] Further, a formal approach does not guarantee success in software development, and is only one possible way of approaching the software development task.[137]

Further, whilst large software development houses such as Microsoft, Adobe, and Oracle might have a prominence in the marketplace, not all software is developed by such large software houses. Often, smaller development teams are involved. Further, as Brooks notes, the "genius" often arises from the work of individual developers.[138] Similarly, it is not the large-scale software architectures which form the basis of patent applications, but instead the granular software components. So a focus on the software product, and the organisation which produces it, overlooks that this is basically a human activity.

When productivity is determined by human factors, it is suggested that aesthetic perceptions over-bears rational analysis.[139] It is possible to set out three ways in which the aesthetic influences underly software development, being those set out by Sinclair in relation to mathematics, and discussed in the Chapter 4 – namely the evaluative, motivational, and generative.

"People and Organizations in Software Production: A Review of the Literature" (1988) 11 (3) *ACM SIGCPR Computer Personnel* 10; Steve McConnell, "10x Software Development" <http://blogs.construx.com/blogs/stevemcc/archive/2008/03/27/productivity-variations-among-software-developers-and-teams-the-origin-of-quot-10x-quot.aspx> (8 June 2011).

134 Meurig Beynon, Russell Boyatt, and Zhan En Chan, "Intuition in Software Development Revisited" in Jim Buckley, John Rooksby, and Roman Bednarik (eds) *Proceedings of 20th Annual Psychology of Programming Interest Group Conference* (2008) <http://www.cs.st-andrews.ac.uk/~jr/papers/ppig08Proceedings.pdf> (16 June 2011), 95 at 96.

135 Beynon et al., above n134 at 96.

136 Beynon et al., above n134 at 96.

137 A number of various approaches, and their problems, are canvassed in Berry, above n115.

138 "A little retrospection shows that although many fine, useful software systems have been designed by committees and built as part of multipart projects, those software systems that have excited passionate fans are those that are the products of one or a few designing minds, great designers.": Brooks, above n11 at 18.

139 "In programming, as in many fields, the hard part isn't solving problems, but deciding what problems to solve. Imagination is hard to measure, but in practice it dominates the kind of productivity that's measured in lines of code.": Graham, above n70.

C Evaluative aesthetic

It was noted above that when implementing a particular algorithm, it is possible to assess the efficiency of code by the extent to which it uses the physical resources of the computer on which it executes. Code that requires fewer CPU cycles to execute would be more efficient than code which does not. But this is not the only criterion by which code might be assessed. The following discussion summarises the possibilities:

> Different programmers will have different definitions for great code. Therefore, it is impossible to provide an all-encompassing definition that will satisfy everyone. However, there are certain attributes of great code that nearly everyone will agree on, and we'll use some of these common characteristics to form our definition. For our purposes, here are some attributes of great code:
>
> - Great code uses the CPU efficiently (that is, the code is fast).
> - Great code uses memory efficiently (that is, the code is small).
> - Great code uses system resources efficiently.
> - Great code is easy to read and maintain.
> - Great code follows a consistent set of style guidelines.
> - Great code uses an explicit design that follows established software engineering conventions.
> - Great code is easy to enhance.
> - Great code is well tested and robust (that is, it works).
> - Great code is well documented.

We could easily add dozens of items to this list. Some programmers, for example, may feel that great code must be portable, must follow a given set of programming style guidelines, must be written in a certain language or must *not* be written in a certain language. Some may feel that great code must be written as simply as possible while others may feel that great code is written quickly. Still others may feel that great code is created on time and under budget. And you can think of additional characteristics.[140]

Some of these attributes clearly contemplate efficiency of use of the physical resources provided by the machine as a goal. These are in the nature of rational, engineering criteria. However other attributes suggest more aesthetic considerations, such as how easy such code is to read and maintain, and its consistency with the social norms which a programming team has adopted. Perhaps the most telling feature however is the admission at the beginning of the quoted passage that different programmers are likely to have different definitions. Code which is efficient may nonetheless be considered "bad" because it is difficult to extend, hard to maintain, or impossible to read.

---

140 Randall Hyde, *Write Great Code: Understand the Machine* (No Starch Press, 2004) at 8–9.

This is a product of the fact that "you can reach the same goal ... by coding in many different styles, using different modules and deploying the same modules in different ways."[141] The subjective nature of the attributes of great code is indicative of the aesthetic nature of *all* of these criteria, since beauty is, after all, in the eye of the beholder. Sinclair notes in relation to mathematics, the aesthetic fulfils an evaluative role because:

> mathematical reality cannot provide its own criteria; that is, a mathematical result cannot be judged important because it matches some supposed mathematical reality—mathematics is not self-organized.[142]

The same is true of software development. Even seemingly rational criteria like the efficient use of resources are arbitrary. As noted above in relation to optimisation, a focus on efficiency may impact negatively on software development, since it adversely effects other criteria such as readability, complexity, and may thereby reduce maintainability.[143]

It has been suggested that aesthetic considerations are also "actively involved in mathematicians' decisions about expressing and communicating their own work".[144] This is clearly also the case in software, where readability, or the related criterion of maintainability is suggested as an important quality of good code.[145]

D   *Motivational aesthetic*

The motivational aesthetic encompasses the motivations which guide the actions of programmers when writing software. As with mathematics, when it comes to programming, it is "difficult to argue that there is an objective perspective – a ... reality against which the value of [software] products can be measured. Contrast this with physics, for example, another discipline that makes strong aesthetic claims ... where questions and products can be measured up against physical reality: How well they explain the shape of the universe or the behaviour of light."[146] The lack of objective criteria against which to make value judgements about software was discussed in relation to the evaluative aesthetic above.

---

141 Stas Bekman and Erik Cholet, *Practical mod_perl* (2003) at 453. As the authors note, this fact is celebrated by the "main motto" of the Perl programming language, the acronym TMTOWDI (there's more than one way to do it).
142 Nathalie Sinclair, "The Roles of the Aesthetic in Mathematical Inquiry" (2004) 6(3) *Mathematical Thinking and Learning* 261 at 265.
143 See the discussion on page 172 above.
144 Sinclair, above n142 at 266.
145 "Programs must be written for people to read, and only incidentally for machines to execute.": Abelson and Sussman, above n32 at xvii. The approach known as literate programming favours readability over all other considerations. See Donald E Knuth, *Literate programming* (Center for the Study of Language and Information, Stanford University, 1992) at 99.
146 Sinclair, above n142 at 274.

Beyond that, the motivational aesthetic concerns notions like the way that an aesthetic stimuli[147] can motivate programmers, either as an indication of the fruitfulness of an avenue of research, affirming the interest in and value of a problem area, and even sustaining a persistent pursuit over a number of years.[148]

At a general level, the motivational aesthetic is to be found in the claims that programming is "fun". Anecdotal evidence supports the claim that software developers, or at least the best software developers, are driven by aesthetic factors, rather than strictly logical ones:[149]

> [t]heir defining quality is probably that they really love to program. Ordinary hackers write code to pay the bills. Great hackers think of it as something they do for fun, and which they're delighted to find that people will pay them for.[150]

Given the empirical evidence which suggests a variance in programmer productivity of an order of magnitude,[151] the importance of the motivational aesthetic is greater than it might otherwise seem. So what makes programming fun? Fred Brooks in *The Mythical Man Month* gives a detailed account of what makes programming fun, which is set out in detail below:

> Why is programming fun? What delights may its practitioner expect as his reward?
>
> First is the sheer joy of making things. As the child delights in his mud pie, so the adult enjoys building things, especially things of his own design. I think this delight must be an image of God's delight in making things, a delight shown in the distinctness and newness of each leaf and each snowflake.
>
> Second is the pleasure of making things that are useful to other people.

147 Sinclair lists the following in relation to mathematics: visual appeal, simplicity, order, "fit", surprise, paradox and the social dimension as evoking a motivational response. See Sinclair, above n142 at 275–277.
148 See Sinclair, above n142 at 274–275.
149 In relation to the Linux kernel, the creator and project leader, Linus Torvalds claims "[t]he most important design issue ... is that Linux is supposed to be fun": in Ann Brashares, *Linus Torvalds, Software Rebel* (Twenty-First Century Books, 2001) at 45. Open source software pioneer Eric S Raymond similarly notes in respect of open source software, "[w]e have *fun* doing what we do ... Our creative play has been racking up technical, market share, and mind-share success at an astounding rate. We're proving not only that we can do better software, but that *joy is an asset*.": Eric S Raymond, *The Cathedral and the Bazaar: musings on Linux and Open Source by an accidental revolutionary* (O'Reilly Media, 2001) at 60. Python language creator Guido van Rossum similarly notes "[i]f there were no art in it, it wouldn't be any fun, and then I wouldn't still be doing it after 30 years.": in John Littler, "Art and Computer Programming" *O'Reilly OnLamp.com*, 30 June 2005 <http://onlamp.com/pub/a/onlamp/2005/06/30/artof prog.html> (25 May 2011).
150 Graham, above n70.
151 See above n133 on page 20 and accompanying text.

Deep within, we want others to use our work and to find it helpful. In this respect the programming system is not essentially different from the child's first clay pencil holder "for Daddy's office."

Third is the fascination of fashioning complex puzzle-like objects of interlocking moving parts and watching them work in subtle cycles, playing out the consequences of principles built in from the beginning. The programmed computer has all the fascination of the pinball machine or the jukebox mechanism, carried to the ultimate.

Fourth is the joy of always learning, which springs from the nonrepeating nature of the task. In one way or another the problem is ever new, and its solver learns something: sometimes practical, sometimes theoretical, and sometimes both.

Finally, there is the delight of working in such a tractable medium. The programmer, like the poet, works only slightly removed from pure thought-stuff. He builds his castles in the air, from air, creating by exertion of the imagination. Few media of creation are so flexible, so easy to polish and rework, so readily capable of realizing grand conceptual structures...

Yet the program construct, unlike the poet's words, is real in the sense that it moves and works, producing visible outputs separately from the construct itself. It prints results, draws pictures, produces sounds, moves arms. The magic of myth and legend has come true in our time. One types the correct incantation on a keyboard, and a display screen comes to life, showing things that never were nor could be.

Programming then is fun because it gratifies creative longings built deep within us *and delights sensibilities we have in common with all men.*[152]

The love of making is something that unites software development with traditional artistic pursuits.[153] In addition, delighting the sensibilities as a motivation adverts to the importance of aesthetics in software development.[154]

---

152 Brooks, above n118 at 7–8 (emphasis added).
153 "What hackers and painters have in common is that they're both makers. Along with composers, architects, and writers, what hackers and painters are trying to do is make good things. They're not doing research per se, though if in the course of trying to make good things they discover some new technique, so much the better.": Paul Graham "Hackers and Painters" <http://www.paulgraham.com/hp.html> (1 August 2011).
154 Hunt and Thomas describe the source of such an emotional response along similar lines to Brooks: "We can create awe-inspiring works with little more than the exertion of the imagination. Why do we do it? We do it for the pleasure of watching them show it off to others, of watching them use in novel ways we'd never imagined. For the thrill of watching millions on millions of dollars in transactions flow through your application, confident in the results. For the joy of building and being part of a team, and for the satisfaction of knowing that you started with a blank canvas and produced a work of art.": Andrew Hunt and David Thomas, "The Art in Computer Programming" September 2001 <http://media.pragprog.com/articles/other-published-articles/ArtInProgramming.pdf> (1 August 2011) at 8. See also Donald E. Knuth, "Computer Programming as an Art" (1974) 17(12) *Communications of the ACM* 667 at 670: "[P]rogramming can give us

The evaluative aesthetic also plays a role in influencing programmers to select the area they work in, and the projects they work on.[155] Within the scope of a particular project, choices also need to be made about the work system they use.[156] Programmers, for whom the computer is the whole of their work environment can be inspired by the aesthetics of the tools they have to work with – text editors, languages, version control systems, libraries, and frameworks.[157] Even visual components such as typography may have a role to play.[158] Tool selection may also be more of a social, than a technological choice.[159] Evidence of the

---

both intellectual and emotional satisfaction, because it is a real achievement to master complexity and to establish a system of consistent rules."

155 "Along with good tools, hackers want interesting projects. ... [A]ny application can be interesting if it poses novel technical challenges.": Graham, above n70.

156 A number of different methodologies are available, each with "a way to manage complexity and change so as to delay and moderate the [time where the software's structure has so decayed that it is very difficult to change anything without adding more errors than have been fixed by the change]. However, each method has a catch, a fatal flaw, at least one step that is a real pain to do, that people put off.": Berry, above n115 at 56.

157 "What do great hackers want? Like all craftsmen, hackers like good tools. In fact, that's an understatement. Good hackers find it unbearable to use bad tools. They'll simply refuse to work on projects with the wrong infrastructure.": Graham, above n70. See also Knuth, above n154 at 672: "[I]t is still a pleasure to do routine jobs if we have beautiful things to work with. ... *Please* give us tools that encourage us to write better programs, by enhancing our pleasure when we do so." Bruce J. MacLennan, "Aesthetics in Software Engineering" (Technical Report No UT-CS-06-579, Department of Electrical Engineering and Computer Science, University of Tennessee, 31 October 2006) <http://web.eecs. utk.edu/~library/TechReports/2006/ut-cs-06-579.pdf> (26 May 2011) at 5, makes relevant observations drawn more generally to all those who use computers, drawing a useful analogy between architecture and software development:

> [F]or many people the computer is not simply one tool in an otherwise uncomputerized occupation; rather, the computer and its software constitute, to a large degree, the entire occupation. In these cases the software system defines the work environment as fundamentally as the physical workspace does. Therefore, the aesthetics of the software systems deserves at least as much attention as that due the architecture, decor, etc. (From this perspective, many contemporary programs are the software equivalent of sweatshops: cluttered, dangerous, ugly, alienating, and dehumanizing.) As architecture deals with the functionality and aesthetics of physical space, organizing it for practicality and beauty, so software engineers organize cognitive (or virtual) space toward the same ends. Thus software aesthetics can have a major effect on quality of work and quality of life.

158 Philip L. Frana, "An Interview with Donald E Knuth", *University of Minnesota Digital Conservancy*, 8 November 2001, <http://conservancy.umn.edu/bitstream/107413/1/oh332dk.pdf> (1 August 2011) at 17–18: "[O]ne of the whole ideas of strutred and literate programming is that you have to be able to understand [the program's] complicated whole. ... With good typography you can perceive the structure, instead of imagining the text as just a chaotic string of characters."

159 "When you decide what infrastructure to use for a project, you're not just making a technical decision. You're also making a social decision, and this may be the more important of the two. ... [W]hen you choose a language, you're also choosing a community.": Graham, above n70. "Aesthetic appreciation can unite a software development organization

motivational aesthetic is also to be found in the impact of other factors such as the workplace,[160] and social factors such as team dynamics.[161]

Even with these preliminary and external factors in place, there are design decisions to be made about algorithms[162] and data structures to use, and how to express them. As in mathematics, the aesthetic may guide these selections. Given the highly complex nature of software creations, rational analysis of the impact of change on a system becomes impossible.[163] Therefore software developers are reliant on aesthetic judgments, simply put, that "designs that look good *are* good".[164]

E   Generative aesthetic

Given the personal, and largely internal nature of much of the activity of programming, there is little direct evidence of the way in which the aesthetic influences the process of inquiry. Undoubtedly this is because we don't understand the way in which the spark of creativity ignites, generally, since it operates at a "tacit or even subconscious level, and intertwined as it frequently is with intuitive modes."[165] But it is suggested that the role of the aesthetic in "the discovery and invention of solutions or ideas; in guiding the actions and choices that [software developers] make as they try to make sense of objects and relations."[166]

In the context of mathematics, Sinclair identifies three strategies mathematicians use for evoking the generative aesthetic: "playing, establishing intimacy, and capitalizing on intuition".[167]

---

through a common set of values embodied in a shared sense of elegance.": MacLennan, above n157 at 4.

160 "After software, the most important tool to a hacker is probably his office. Big companies think the function of office space is to express rank. But hackers use their offices for more than that: they use their office as a place to think in. And if you're a technology company, their thoughts are your product.": Graham, above n70. See also Tony DeMarco and Tim Lister, "Programmer Performance and the Effects of the Workplace" in *Proceedings of the Eighth International Conference on Software Engineering*, Longon, August 1985, 268; Nash and Redwine, above n133 at 14.

161 See Nash and Redwine, above n133 at 14–15; Brooks, above n11 at 18.

162 "Given a solvable problem, there are many algorithms (programs) to solve it, not all of equal quality. The primary practical criteria by which the quality of an algorithm is judged are time and memory requirements, accuracy of solution, and generality.": *Encyclopedia of Computer Science*, (Wiley & Sons, 2003) <http://www.credoreference.com/entry/5880599> (27 July 2011). It should be noted however, despite the seemingly technical choice just described, that algorithms are not often chosen like off-the-shelf components, instead the algorithm slowly emerges from repeated attempts of the programmer to describe the method in computationally descriptive form. This is what Brooks means when he says that software is "grown, not built.": Brooks, above n11 at 18.

163 MacLennan, above n112 at 5.

164 MacLennan, above n112 at 3–4.

165 Sinclair, above n142 at 270.

166 Sinclair, above n142 at 270.

167 Sinclair, above n142 at 271.

Play: Sinclair notes the role of "free play" in allowing mathematicians to develop their craft. "[T]he mathematician, freed from having to solve a specific problem using the analytical apparatus of her craft, can focus on looking for appealing structures, patterns and combinations of ideas."[168] Some programmers suggest a similar ploy for developing proficiency, through the writing of "toy programs".[169] The camel code in Listing 5.1 above is a good example of such a program. On a similar note, programmers involve themselves in open source software development projects beyond paid work in order to develop their proficiency. Other forms of experimentation include participation in obfuscated code contests, and the development of "weird" languages.[170]

In relation to particular projects, the early stages of a project may involve free play in a different form, by simply doodling on paper, exploring structures and concepts in order to come to an understanding of the nature of the project domain. Some approaches to software development, such as rapid prototyping and incremental release methodologies such as agile and XP, also have this sort of aesthetic experimentation at their core, since they encourage building code as a way of exploring a domain, and generating aesthetic feedback through use, or experience, of the software from an early stage. More formal approaches such as requirements engineering, and structured programming, use other written works such as formal specifications as the mechanism of exploration.

Personal, intimate relationships: Sinclair also draws attention to the way mathematicians become familiar with an area by naming the objects they are dealing with, with a view to getting "some traction on still vague territory".[171] There is some evidence of this sort of anthropomorphisation by programmers.[172] However, the development of a relationship between programmer and code is centred around the development of a prototype. This desire to develop the coder-code relationship explains the "code first, then debug" approach adopted by many novices, and even some experienced programmers.[173] Developing a relationship with the domain is also at the heart of more modern iterative approaches such spiral development, agile, extreme programming.[174] Although other more formal, structured methodologies don't involve an early relationship with code

---

168 Sinclair, above n142 at 272.
169 Knuth, above n154 at 672. Dave Thomas, advocates a similar approach, which he calls "Code Kata": <http://codekata.pragprog.com/codekata/2007/01/code_katahow_it. html> (11 June 2011).
170 See Thomas Taylor, "Obfuscation, Weird Languages and Code Aesthetics" on *obfuscators. org*, 26 April 2008 <http://www.obfuscators.org/2008/04/obfuscation-weird-lan guages-and-code.html> (1 August 2011).
171 Sinclair, above n142 at 272.
172 See for example Eric S. Raymond, "Anthropomorphization" in *The Online Hacker Jargon File* <http://www.catb.org/~esr/jargon/html/anthropomorphization.html>.
173 See Berry, above n115, at 51–53; Boehm discusses the evolution of the hacker culture, which lauds such cowboy programmers in Boehm, above n113 at 13; Paul Graham asserts a personal view that this is the only way to develop software: Graham, above n153.
174 These methodologies, and their weaknesses are discussed in Berry, above n115 at 58–65.

in a strict sense, requirements documentation, formal analysis, requirements engineering, are geared towards building familiarity with the concepts that lie behind the code, and may therefore be treated as involving a similar process.[175]

The need for a relationship with the code is also behind one of the great drivers of change in software projects, the "IKIWISI" phenomenon,[176] whereby the early prototype informs not only programmers' conception of the software being created, but also the client's or end user's.[177] This in part explains the drive to change software throughout its development.

Finally Brooks' notion that software should be grown, not built, is a reflection of "an organic process of interaction between the embryonic software product and its environment that takes place in the developer's mind."[178]

Capitalising on intuition: Peter Naur, an important figure in computer science, advanced a position in the 1984 that intuition was "the basis on which all activites involved in software development must build."[179] Despite Naur's status as a "most distinguished contributor to the study of software practice, and ... a receipient of the Turing Award in 2005",[180] the paper has received only a handful of references in subsequent academic literature.[181] Adherents to a strict, formalistic, or rationalist approach to software development are likely to equate intuition with guesswork, or a lack of risk aversion.[182] So is it relevant to suggest a central role for intuition in mathematics?

Poincaré characterised the mathematical generative process in mathematics as being concerned with, not the logical, but "the construction of possible combinations of ideas and the selection of the fruitful ones."[183] As a matter of intuition, experience,[184] and on the basis of the Curry-Howard isomorphism, it is asserted that software development is similarly concerned with the combination and selection of such ideas. The aesethetic assists in the assessment of combinations of ideas which are "harmoniously disposed so that the mind can effortlessly embrace their totality without realising their details."[185] The intangible nature of

---

175 See Berry, above n115 at 58–61.

176 This stands for "I'll know it when I see it".

177 In this author's experience, and humble opinion, it often seems to be the case that clients don't really have any conception of what they want, but are quite adept at explaining how a prototype is *not* what they want.

178 Beynon et al., above n134 at 99.

179 Peter Naur, "Intuition in Software Development" in Harmut Ehrig et al. (eds) *Theory and Practice of Software Development*, Volume 2, LNCS 186 (Springer, 1985) 60. For a more recent update, see Beynon et al., n134.

180 Beynon et al., above n134 at 95.

181 Beynon et al., above n134 at 95.

182 "One of the main purposes of adopting a formal approach to program development is to shift the focus from 'considering your intuitive ideas about what a program is intended to do' to 'thinking precisely and abstractly about what is it [*sic*] that you have instructed it to do'." Beynon et al., above n134 at 96 (citations omitted).

183 Sinclair, above n142 at 270.

184 See also the comments of Charles Strauss in Philip J Davis and Reuben Hersh, *Descartes' Dream: The World According to Mathematics* (Dover Publications, 1986) at 180.

185 Sinclair, above n142 at 270.

software[186] means that the mind's traditional tools for dealing with spatio-temporal objects are not available.[187] The complexity of software creations means that it is very difficult to rationally assess the effect of software design decisions or changes on the whole of software.[188] Ershov aptly captures the complexity of the programming task, which he describes as "the most humanly difficult of professions",[189] because:

> In his work, the programmer is challenged to combine, with the ability of a first-class mathematician to deal in logical abstractions, a more practical, a more Edisonian talent, enabling him to build useful engines out of zeros and ones, alone. He must join the accuracy of a bank clerk with the acumen of a scout, and to these add the powers of fantasy of an author of detective stories and the sober practicality of a businessman. To top all this off, he must have a taste for collective work and a feeling for the corporate interests of his employer.[190]

In other words, "[t]he scale of software development is such that no single designer can appreciate all the perspectives that are relevant to the effective solution of problems."[191] Therefore it is only through aesthetic impressions that the software design and construction process is informed during the generative phase. Indeed it may be that Brooks' formulation of "conceptual integrity" as a paramount consideration in sucessful software projects is referable to the an intuitive understanding of the project in the mind of the project leader.[192] A recognition of the aesthetic in programming demands "complete recognition and full exploitation of the broad scale of individual activities in programming."[193]

186 "The essence of a software entity is a construct of interlocking concepts: data sets, relationships among data items, algorithms, and invocations of functions": Brooks, above n11 at 11.
187 Brooks, above n11 at 12.
188 "In most cases, the elements interact with ech other in some nonlinear fashion, and the complexity of the whole increases much more than linearly.": Brooks, above n11 at 11.
189 Ershov, above n124. See also Brad Cox, "No Silver Bullet Revisited" *American Programmer Journal*, November 1995 <http://virtualschool.edu/cox/pub/NoSilverBulletRevisted> (22 July 2011): "[P]rogrammers invariably agree that computer software is the most complicated of all human activities. I've been heard [*sic*] word processors seriously compared with the complexity of Boeing's airplanes."
190 Ershov, above n124 at 502. Ershov lists a further two reasons why software creation is such a complex task, namely that their work "brings them to those limits of human knowledge which are marked by algorithmically unsolvable problems and which touch upon deeply secret aspects of the human brain": *ibid*, and that the programmer's personal push-down stack "must be as deep as is needed for the problem, plus at least 2–3 positions deeper": *ibid*. A personal push-down stack is a reference to the ability to juggle tasks in a last-in-first-out fashion. He asserts that the average person has a stack depth of 5–6 tasks. See also Letter from Andrei P. Ershov to Frederic L. Coombes, 18 November 1972 <http://ershov.iis.nsk.su/archive/eaindex.asp?lang=2&did=382> (21 June 2011).
191 Beynon et al., above n134 at 98.
192 Beynon et al., above n134 at 99.
193 Ershov, above n124 at 505.

F Summary

The foregoing discussion underscores the creative nature of the programmer. Aesthetic considerations influence every aspect of software development, including: assisting in the evaluation of code and infrastructure quality; guiding project selection and influencing productivity; and guiding the activities of individual developers. The aesthetic also features in the way that programmers use experimentation and intuition to manage complexity, perceive solutions, and advance the state of the art. Software development does not proceed by a sequence of logical steps, but, as with any creative endeavour, involves a winding, internal, and seemingly irrational, path from concept to code.

2.4 Programming is not in the useful arts

To review, the results of the above three-dimensional analysis are as follows. First, it is entirely possible to analyse claims to software on a physical level. The physicality of software has been located by courts in the computer hardware on which software eventually executes, and the medium on which software is stored before execution. But there is a significant difference between software as written, and software as executing. To focus on the end product, agonising about what software is, and to point to a particular physical device as being the locus of software's existence is to ignore the aspect of software which is most significant to its creation. As far as programmers (and software's users) are concerned, it is the symbolic aspect of software which is most important.

This being the case, a claimed invention's abstractness or physicality needs to be very carefully considered to determine whether the physical aspects of the computer hardware, or associated peripherals, impose any meaningful limitation on its development. Another way of putting it is to ask whether the software is merely a component of a larger device (not being a general purpose computer) or whether what is claimed is an abstract artefact, or a physical invention. Artefacts occuring higher up the stack, such as algorithms and abstract data structures are sufficiently abstract that they ought never to be considered patentable. As noted in Chapter 1, the history of software development suggests that, in general, software will be crafted at higher and higher levels of abstraction. Even at the present time, the majority of software is written at a level of abstractness beyond the realm of meaningful physical limitations.

Second, software implements a function, namely a computable process. However, the expressiveness of software remains a significant part of the software development process. It permeates every level of the abstraction stack, and is not stripped from software even by its translation to executable form where it contains, at least a kernel of technical expressiveness. Beyond that, it is submitted that executable software is expressive in other ways. Software communicates to users the way in which it is to be used to achieve particular goals. Machine code also has an informative aspect, in that it continues to describe a particular computable process, which, with the benefit of decompilation tools, can be extracted. It will only really be where the physical aspects of computer hardware significantly restricts the number of ways in which a particular process can be implemented that the expressiveness of software is curtailed. As software continues to grow more and more abstract, software development is likely to become more and more expressive.

The foundation in logic and formal mathematics, as well as the business-like nature of the enterprise, and the desire to make software into a form of engineering, all create an impression of software development as a rational activity. But software is not created by progressing through a sequence of logical steps, rather by travelling a winding and tortuous path. Software development is a creative, artistic pursuit, governed by aesthetic considerations. Aesthetics determines the criteria by which software is adjudged good or bad; draws programmers towards particular development projects; motivates developers to create; and is key to the advancement of the discipline through the generation of innovative approaches to development. Although the development of software may be assisted by structured methodologies, they cannot address the essence of software development, only its accidents, and are therefore only of secondary importance. This will be the case irrespective of the level of physicality of the claimed invention, since in all cases, the activity of programming is directed to crossing the semantic gap from a human understanding of desired function, to a complete and precise description of a computable process.

Taking all this into account, it is clear that the creation of software, in all but exceptional circumstances, falls within the fine arts, rather than the useful arts. Undoubtedly there will be occasions where the factual matrix of a particular set of claims ought to be considered very carefully, before a final decision is made. But it is where the distinctions are the slightest, that the most resolute adherence to such distinctions is required. The framework just used holds value as a way of analysing a particular set of claims, to determine how physical, how expressive and how aesthetic the development process is likely to be, and whether therefore such a claimed invention is patentable subject matter. But the general analysis just completed makes clear that the starting point for any claims to software should be that software is *not* patentable. This being the case, a categorical exclusion for software is appropriate.

Where what is claimed involves a software component, but the creation of such software is dominated by physical, functional, and rational considerations, then it might be concluded that the essence of that invention is not the software which is claimed, but the physical hardware, and that as such the invention falls outside of a categorical exclusion. Such a position is consistent with the claimed isomorphism between mathematics and software, since the "software" in such instances is a mere description of a component of a physical device, which accords with the orthodox view that a mathematical description of the parts of a physical invention would not be a bar to patentability.

3 Conclusion

This chapter has looked at the patentability of software, as determined by the characteristics of the activity involved in creating it, namely programming. There is no doubt that software falls near the borderline of patentable subject matter. But for the reasons set out above, it is clear that it falls outside the realm of patentable subject matter. When what is claimed is the product of programmers' creative endeavours, it will in most cases be an intangible, expressive, and aesthetic artefact which is claimed, and is not within the field of the useful or technological arts.

The contention that software is not patentable subject matter has clear implications, both for the software industry, the patent regime, and intellectual property more broadly. It is to those implications that the final chapter of this thesis is devoted.

6 Implications

1 Introduction

The last chapter demonstrated why programming, like mathematics, is not a useful art, with the consequence that software ought not be patentable. This chapter explores the implications of that position. First of all, the bounds of subject matter as presently understood in the three jurisdictions considered in this thesis are considered in light of the analytical framework adopted.

Although software development is a fine art, rather than a useful art, simply declaring that software is not patentable is unlikely to be dispositive of the issue. The history of the exclusion of computer programs "as such" under the European Patent Convention stands as a testament to that. With the growing ubiquity of computers, software components have become a part of many industrial processes, and components of all manner of common physical devices, from watches to cars. These products are part of the traditional domain of patent law. Therefore, there is at the very least a need to distinguish claims to a patentable invention containing a computable process, and non-patentable claims to the process itself. This stresses the importance of properly characterising the invention.

Next, the role of subject matter is considered. The need to exclude software from the patent paradigm suggests that subject matter is not a "failed gatekeeper" which should be retired. Some consideration is also given to the role of freedom in patent law. Whilst the relationship between copyright and free expression is much better travelled, the role of freedom in patent law is much less discussed. There has been some consideration given to the influence of human rights jurisprudence on intellectual property, and in particular patent law.

Having considered the implications of the *prima facie* non-patentability of software on the patent regime, some remarks are addressed to a broader issue. As it is suggested that software ought not to be protected within the patent regime, what are the implications of such a proposal on the protection which software should, would, or perhaps already does receive?

Because this is the final chapter, and is followed by the conclusion, it does not have its own concluding section.

2 An assessment of the framework

At the outset of the thesis, three features of software which make it special were set out: abstractness, complexity, and reuse. A few comments are now directed towards how the framework developed and applied takes account of those features.

The level of abstractness of a particular invention is clearly assessed in the physicality versus intangibility dimension of the framework. The more abstract a piece of software is, the less connection it has to the workings of the physical computer on which it runs. The abstractness of software also affects the available expressivity since, as asserted in Chapter 5, it is only at very low levels of abstraction that the physical limitations of hardware will restrain the ability to express a computable process in more than one way.

The complexity of software was explored in the third dimension of the framework, the aesthetic versus the rational, and it was noted that the complexity of software means its construction cannot depend on a rational analysis of its states. As such, the complexity of software results in a reliance on aesthetic responses, or intuition, to guide the creative process. The framework clearly addresses this aspect of software's nature.

The need for reuse is related to all three limbs of the analytical framework. It is the intangibility of software which makes its reuse simple. Aesthetics depend on and are guided by previous experience. A dependence on reuse also suggests the need for freedom *to* reuse, and as such lies at the foundation of the analytical framework. To look at it another way, a culture of reuse is a hallmark of a fine art.[1]

3 An assessment of the state of patent law

Given the argument put forward in Chapter 5, that software ought not be patentable, it is worthwhile to compare how the current state of subject matter in the three jurisdictions considered in this thesis fares. All three jurisdictions considered are signatories to the TRIPS treaty,[2] Article 27 of which requires that patents be made available in all fields of technology. So on that basis, the three jurisdictions should be compatible with the analytical framework, since the framework determines what is in the field of technology as against what is outside it.

1 An illustration of the importance of reuse to music is made at the end of the movie *Patent Absurdity: How Software Patents Broke the System* (Directed by Luca Lucarini, 2010) <http://patentabsurdity.com/> (27 August 2011) at 27:45. See also Richard Stallman, "The Danger of Software Patents," Transcript of Speech Given at Cambridge University, March 2002, <http://www.cl.cam.ac.uk/~mgk25/stallman-patents.html> (27 August 2011): "Beethoven, as it happens, had a lot of new musical ideas but he had to use a lot of existing musical ideas in order to make recognizable music."

2 *Marrakesh Agreement Establishing the World Trade Organization*, opened for signature 15 April 1994, 1867 UNTS 3, annex 1C (*Agreement on Trade-Related Aspects of Intellectual Property Rights*) (entered into force 1 January 1995).

3.1   The United States

The United States limitation of patent law to the useful arts is obviously consistent with the analytical framework proposed. The current understanding of that test recognises three judicial exceptions to the four categories of patentable subject matter in the US Constitution, namely "laws of nature, physical phenomena, and abstract ideas",[3] and relies on the machine-or-transformation test a "useful and important clue" as to its patentability.[4] The machine-or-transformation test is directed to the physicality dimension of the three dimensions used in this thesis. It was noted how the expressiveness of software can depend at least to some degree on the physicality analysis, so there is at least some indirect consideration of that dimension. This expressiveness was directly raised in the *Prater* rehearing,[5] but then the case was decided on other grounds. This could have been characterised as an avoidance of the issue, or a downplaying of its significance.

The doctrine of mental steps[6] was theoretically directed to both the abstract nature of the process claimed. But the failure of the doctrine was not, as argued by Coulter, that it did not discriminate between technological, rational processes which could be carried out without human intervention, and those including "peculiarly human mental activities which cannot, in principle be performed by devices".[7] The failure of the doctrine was that it did not consider the other two dimensions of the analysis used, which are concerned with the nature of the creative process. Similar problems may attend on the "abstract ideas" exclusion, if it is said that ideas are patentable only because they are not physical.[8] The machine-or-transformation test is on its face directed towards only one dimension of analysis as well. As was argued in Chapter 5, the physical limitations of computer hardware might reduce the expressivity of software, but say nothing about the aesthetic versus rational nature of its creation.

---

3 *Alice Corp v CLS Bank International* 134 S. Ct. 2347 (2014) (*"Alice Corp"*) at 2355; *Mayo Collaborative Servs v Prometheus Labs, Inc* 132 S. Ct. 1289 (2012) (*"Mayo"*) at 1293.

4 *Bilski v Kappos* 130 S. Ct. 3218 (2010) (*"Bilski"*) at 3226 per Kennedy J, 3258 per Breyer J; *Mayo* at 1303.

5 *In re Prater* (1969) 415 F.2d 1390. The expressiveness was raised in the form of an objection that fell foul of the First Amendment right of freedom of thought.

6 Discussed in Chapter 2, at 40.

7 Robert I. Coulter, "The Field of the Statutory Useful Arts: Part I" (1952) 34 *Journal of the Patent Office Society* 425 at 426. Coulter's approach conflates the nature of the process claimed, with the nature of the activity involved in producing it. As has been argued in this thesis, just because a process, for example, a computable process, is defined, and even carried out, as a sequence of logical steps, it does not follow that the creative process behind it is similarly logical or rational. See Chapter 5.

8 In the decision in *CyberSource Corporation v Retail Decisions Inc* Appeal No 2009-1358 (Fed Cir, 2011) (*"CyberSource"*), Dyk J held the claimed software method non-patentable as a mental process, a subcategory of the abstract ideas exception, because it "can be performed in the human mind, or by a human using a pen and paper": at 12. Such an approach seems to reinvigorate the mental steps doctrine, in an albeit narrower way, since claims which might be performed by a human might yet be patentable if the claims are limited by a machine or transformation.

The majority approach in *Bilski* is more problematic. Proper distinctions between the useful arts and the fine arts cannot be made by asking whether something is a process within the "ordinary, contemporary, common meaning"[9] of the word. As Stevens J correctly noted,[10] the supposed literal interpretation of the majority in that case is not able to be reconciled with the specific exclusions for "laws of nature, physical phenomena and abstract ideas".[11] In addition, the ordinary meaning of the word process says nothing about the fact that both fine and useful arts may involve processes, but it is only the latter which are traditionally patentable. To address that distinction it is necessary to consider the nature of the process by which an allegedly *patentable* process is "made". It is urged that a full three dimensional consideration of the bounds of patentability in that jurisdiction is not precluded by *Bilski*, but would requires a change in approach to interpreting the category of patentable processes, one steeped in a historical, purposive interpretation of the constitutional clause, rather than a literal one. This was the approach adopted by Stevens J, although admittedly it represents a minority view.

The exclusion of "laws of nature, physical phenomena, and abstract ideas",[12] does however provide a possible way out of this conundrum. It is perhaps for this reason that subsequent cases have emphasised these exclusions as the starting point in a two-step analysis of patentability.[13] As noted above, the abstract ideas exclusion might only be considered directed towards the intangible versus physical limb of the framework. However, recalling the relationship between thought and expression discussed in Chapter 4,[14] such ideas are better characterised as giving rise to expressive rather than purposive considerations. As the process of creating ideas is a peculiarly human process, and one which is not well understood, such abstract ideas are better described as aesthetic than rational. The nature of mathematics, as primarily grounded in thought and expression, illustrates why the creation of such ideas is the domain of the fine arts, rather than the useful arts.

Similarly, the laws of nature exclusion can be reconciled with the analytical framework used herein. On the basis of the characterisation of mathematics as "the queen of the sciences"[15] and the acknowledged aesthetics of other sciences,[16] it is submitted that a distinction might be drawn between science and

---

9   *Bilski* at 3221 per Kennedy J.

10 *Bilski* (2010) at 3235.

11 *Bilski* at 3221 per Kennedy J, quoting *Diamond v Chakrabarty* 447 US 303 at 309.

12 *Bilski* at 3221 per Kennedy J.

13 *Mayo* at 1296–1298; *Alice Corp* at 2355.

14 See Section 3.1 on page 127.

15 Carl Friedrich Gauss, cited in Guy W Dunnington, Jeremy Gray and Fritz-Egbert Dohse, *Carl Friedrich Gauss: titan of science* (MAA, 2004) at 44.

16 Nathalie Sinclair, "The Roles of the Aesthetic in Mathematical Inquiry" (2004) 6(3) *Mathematical Thinking and Learning* 261 at 274, describing physics as another discipline with "strong aesthetic claims", and citing in support Graham Farmelo (ed), *It must be beautiful: Great equations of modern science.* (Granta Books, 2002).

technology in that the former has strong aesthetic and expressive characteristics, putting it closer to the fine rather than the useful arts.

3.2   The United Kingdom

The relationship between the analytical framework and the EU model is perhaps the most challenging to clearly resolve. On the one hand, the tests used by the EPO and UKPO resolve the enumerated exclusions to a single technicality requirement (whether that be technical contribution or technical effect). Technicality could be taken to be a synonym for technology, and therefore compatible with the analytical framework. However, the resolution of the Article 52(2) categories to a requirement of technical character rests on a dubious legal foundation. Such a formulation of patentable subject matter was explicitly "rejected by the framers of the EPC from the outset".[17] Further, as noted in *Symbian*,[18] the reliance on the concept of technicality in the EPC context is problematic because Article 52 makes "no reference to any 'technical' requirement".[19] As such, in the absence of clear guidance as to the meaning of this concept, such a test will have "all the disadvantages of the original obscure wording, with the added disadvantage of not even providing the actual legislative test".[20]

It may be recalled from Chapter 2, that Article 52(2) of the European Patent Convention declares that the following are not patentable inventions "as such":[21]

(a)   discoveries, scientific theories and mathematical methods;
(b)   aesthetic creations;
(c)   schemes, rules and methods for performing mental acts, playing games or doing business, and programs for computers;
(d)   presentations of information.

But these exclusions share a common characteristic that they are "clearly nontechnical".[22] Two of those specific exclusions, namely computer programs, and mathematical methods, have been analysed according to the proposed framework. The remaining exclusions would, or a cursory analysis at least, seem consistent with that analysis as well. So to that extent, it seems fair to suggest that the analytical framework both fleshes out the features which make the specific

17 Justine Pila, "Dispute over the Meaning of 'Invention' in Article 52(2) EPC – The Patentability of Computer-Implemented Inventions in Europe" (2005) 36 *International Review of Intellectual Property and Competition Law* 173 at 185.
18 *Symbian Ltd v Comptroller-General of Patents* [2008] EWCA Civ 1066; [2009] RPC 1 ("*Symbian*").
19 *Symbian* at [29].
20 *Symbian* at [30].
21 The "as such" proviso is found in Article 52(3).
22 Brad Sherman, "The patentability of computer-related inventions in the United Kingdom and the European Patent Office" (1991) 13(3) *European Intellectual Property Review* 85 at 93.

exclusions non-technical, and may be of benefit in determining whether a particular invention falls within or without a specific exclusions.

3.3   Australia

In Australia, the original manner of manufacture test from the *Statute of Monopolies* remains in force, as interpreted in *NRDC*.[23] The useful arts versus fine arts distinction was drawn in that case, and illustrates a consistent position with the US. The heavy borrowing of guidance from the US on subject matter approaches also is consistent with that position. A 2010 review of the patentable subject matter test by the Advisory Council on Intellectual Property recommended a change to the test, "using clear and contemporary language that embodies the principles of inherent patentability as developed by the High Court in the NRDC case and in subsequent Australian court decisions".[24] No specific wording of that proposed codification was endorsed, although in the discussion, ACIP noted that "subject matter must relate to the useful, rather than the fine, arts".[25]

Physicality was a key component of the *NRDC* decision, and the most recent cases of *Research Affiliates* and *RPL Central* have, as discussed in Chapter 2,[26] continued this requirement.

As is to be expected, given the way that Australian law has tracked the US position, little consideration of the other dimensions of analysis is evident. At best, it may be possible to read a consideration of expressiveness into the traditional exclusions, such as the "scheme or plan", "intellectual information", and "presentation of information" exclusions. As to the third dimension, aesthetics, there is nothing in the law as it currently stands in this jurisdiction which supports this limb of the analysis.

3.4   Summary of the jurisdictions

It is clear from the current state of patent law in all three jurisdictions, that at best, only half of the framework is being considered. In the EU, under the "any hardware" approach, the framework is almost completely ignored.

Some commentators are particularly dismissive of *Bilski*'s machine-or-transformation test.[27] The Supreme Court was itself at least wary of the test, in that it did not endorse it as the sole test of patentability. Yet, as has been argued in this

---

23 *National Research Development Corporation v Commissioner of Patents* (1959) 102 CLR 252.
24 Advisory Council on Intellectual Property, *Patentable Subject Matter* (Final Report, December 2010) ("ACIP 2010") at 13.
25 ACIP 2010 at 13.
26 See page 53.
27 Mark A. Lemley et al. "Life After *Bilski*" (2011) 63 *Stanford Law Review* 1315 at 1316; Dennis Crouch and Robert P. Merges, "Operating Efficiently Post-*Bilski* by Ordering Patent Doctrine Decision-Making" (2011) 25 *Berkeley Technology Law Journal* 1673 at 1690.

thesis, some form of physicality requirement has informed the scope of patent law since its inception. Physicality has also been of particular value in the software patent debate, as it provides a rational framework upon which to draw distinctions between "pure" software patents which would undermine free access to knowledge on the one hand (like the Macrossan invention),[28] whilst still allowing claims to physical devices and methods which include computable processes (like the Aerotel claims)[29] to be patentable.

Therefore it is to be expected that the physicality dimension has been the primary one for patent law, to this point. However, the other two dimensions,[30] have been shown to be particularly useful in that they move beyond an exploration of the nature of the thing claimed, to look at the nature of the process involved in creating it. This change in focus was demonstrated to be key to understanding the reason why mathematics was not patentable, since the competing claims as to the nature of mathematics amount to something of a dead end. Similarly, these two aspects when considered in the context of software defied what might be considered a common understanding of the nature of software, and demonstrated why programming was more an artistic than an engineering activity. As such, the framework illustrates a shortcoming in current approaches to patentability. Confusing ontological issues might be replaced, or at least complemented, by a consideration of the creative process, to determine the appropriateness of the subject matter for the award of patent rights.

4 Broader implications of the argument for patent law

4.1 Subject matter is not a "failed gatekeeper"

The subject matter inquiry is a complex one. It involves a multitude of hard to reconcile cases, many and changing "investigative tool[s]"[31] and a number of historical exclusions. Determinations of what is patentable and what is not depend on a comprehensive understanding of the nature of particular fields of technology as well as the operation of particular inventions. Unsurprisingly then, it may seem like a good solution to take a broad view of eligible subject matter;[32] reduce

---

28 *Aerotel Ltd v Telco Holdings Ltd and in the Matter of Macrossan's Application* [2006] EWCA Civ 1371 (*"Aerotel"*) at [58]–[74].

29 *Aerotel* at [50]–[57].

30 Expressive or functional; aesthetic or rational.

31 *Bilski* at 3227.

32 This is the effect of "useful result" approach, adopted in *State Street* in the US, and *Catuity* in Australia. As Lemley et al. suggest, in the wake of the *State Street* decision, "patentable subject matter was effectively a dead letter": Lemley et al., above n27 at 1318. A broad view is often indirectly adopted when appeal is made to the need for flexibility, or to adapt (or remove) traditional limitations in the wake of new "technologies": see for example *Bilski* at 3227–3228 per Kennedy J. Whilst such a view is consistent with the broad view of technology as discussed in Chapter 4, at 152, both as a matter of intuition and as a matter of precedent, all human activity which advances the sum of human knowledge is not correctly described as "technology".

subject matter to a mere form requirement;[33] expressly disavow the subject matter inquiry in favour of the other technical requirements,[34] narrow the subject matter inquiry to a narrow technical one;[35] or avoid the issue by determining subject matter issues as a matter of last resort.[36] Whilst Lemley et al. are right to draw attention to the difficulties attendant on drawing bright lines between patentable and non-patentable subject matter, "[t]wilight does not invalidate the distinction between night and day".[37]

The argument as to why software ought not be considered patentable need not be restated again here. However, the need for subject matter exclusions is also evident when alternative solutions as to the software patent problem are considered. In light of the arguments advanced in the last chapter, the suitability of these reforms as solutions to the software patent problem are considered below.

A Alternative solutions are insufficient

Stricter disclosure requirements: By enforcing the disclosure requirement much more strictly, it is said the scope of patent claims could be considerably narrowed. It is clear from the analysis in Chapter 7 that to truly describe the invention, the description of a computable process must be at a much lower level down the waterfall than at the design level (that is algorithms and data structures). Firstly, the physicality of the invention needs to be evident in order to satisfy the physical/intangible limb – it should be clear how this physicality limits the claims. But to properly describe a computable process, and in recognition of the need to cross the semantic gap, it is at least arguable that there should be disclosure of source code, or perhaps pseudocode.[38]

On the positive side, such an approach might limit the effective life of the patent to a period consistent with the market life of software innovations. A more extensive disclosure of the invention could also result in better quality documentation of the prior art in the field, and increase the likelihood of patent literature becoming a useful source of technical information for programmers.

One problem with narrowing patent scope however, is that with computer programs, it is "quite possible to produce functionally indistinguishable program behaviours through the use of more than one method".[39] Thus such a solution replaces an over-protection problem with one of under-protection.

33 This is the approach adopted in the European Patent Office, wherein any claim to hardware is sufficient to pass muster under Article 52 of the European Patent Convention.
34 See for example Michael Risch, "Everything is Patentable" (2008) 75 *Tennessee Law Review* 591.
35 Lemley et al., above n27.
36 Crouch and Merges, above n27.
37 Murray Gleeson, "Judicial legitimacy" (2000) 20 *Australian Bar Review* 1 at 8.
38 Pseudocode is an informal way of describing the algorithm behind a computer program independent of the syntactic constraints of a programming language. See for example Wikipedia, "Pseudocode" <http://en.wikipedia.org/wiki/Pseudocode> (8 August 2011).
39 Pamela Samuelson et al., "A Manifesto Concerning the Legal Protection of Computer Programs," (1994) 94 *Columbia Law Review* 2308 at 2345.

But the source code disclosure route must ultimately fail because it does not answer the criticism that programming involves a fine art, a creative process, and is not the sort of activity likely to be incentivised by the grant of patent rights.

Improved examination procedures: Other solutions focus on the possibility of improving patent quality by improving the patent examination process. There have been moves in the US to improve funding to the USPTO[40] and address the high attrition rate of patent examiners.[41] Given the ever-increasing presence of software in daily life and its accompanying growth in complexity, one should wonder whether patent office budgets can keep up. Certainly, despite some advances in the last 10 years, the backlog is still significant.[42]

In 2005 the USPTO implemented a system of peer review of patent applications as a way of providing greater scrutiny by creating "a peer review system for patents that exploits network technology to enable innovation experts to inform the patent examination procedure."[43] Whilst such a reform is a positive step towards filling the gap in the documented prior art base, it is not without problems. One commentator voiced concerns about the ability of such a system to

40 Sarah L. Stirland, "Bush makes new push for patent office to keep fees" *GovExec.com*, 6 February 2006, <http://www.govexec.com/story_page.cfm?articleid=33317> (15 August 2011). On the efforts to manage the backlog of applications, see Courtenay Brinckerhoff, "USPTO Backlog Update: One Step Forward, One Step Back" on *PharmaPatents* <http://www.pharmapatentsblog.com/patent-office-practice/uspto-backlog-update-one-step-forward-one-step-back/> (16 August 2011); United States Patent and Trademark Office "Backlog Stimulus Reduction Plan" <http://www.uspto.gov/patents/init_events/PatentStimulusPlan.jsp> (17 August 2011). The current size of the backlog stands at around 700,000 applications: United States Patent and Trademark Office, "July 2011 Patents Data, at a Glance" <http://www.uspto.gov/dashboards/patents/main.dashxml> (15 August 2011).

41 Olsen F., "Patent examiners battle stress," *Information Today*, 25 July 2005, <http://fcw.com/articles/2005/07/25/patent-examiners-battle-stress.aspx?sc_lang=en> (15 August 2011).

42 As at July 2011, the backlog was 689,226 pending applications, with the average time from filing to a "first action" by the USPTO being 27.8 months, and a total average lag between filing and resolution of 33.5 months: USPTO, "July 2011 Patents Data, at a Glance" <http://www.uspto.gov/dashboards/patents/main.dashxml> (27 August 2011). Looking at the average filings since April 2011, it can be calculated that during that period, over 7,000 applications were filed each week. By July 2017, the total backlog was down to 542,840 pending applications, with the average time from filing to a first action being 16.4 months, and total time to resolution at 24.8 months. There had been a decline in business method patent filings post-*Alice* from around 1200 per month, to around 900 per month. Pendency remains an issue in this category, with a first-action pendency of around 27.6 months, and a total pendency average of 35.4 months. See USPTO, "July 2017 Patents Data, at a Glance" <http://www.uspto.gov/dashboards/patents/main.dashxml> (6 August 2017).

43 "About Community Patent," The Community Patent Project, <http://dotank.nyls.edu/communitypatent/about.html> (4 June 2006). For more detail, see Beth Noveck, "'Peer to Patent': Collective Intelligence and Intellectual Property Reform" *NYLS Legal Studies Research Paper* No. 05/06-18, 25 April 2006, <http://ssrn.com/abstract=898840> (23 August 2006).

deal with the huge volume of patent applications.[44] Based as it is on social software such as Wikipedia,[45] any such system must also deal with similar concerns as to the potential for fraud and bias within such a system, particularly since patent applicants have more at stake than a favourable review in an online encyclopaedia.[46] It is also relevant to note that such a system is national in nature and thus depends on having a sufficient number of suitably qualified expert volunteers with a country's borders, or else relying on people from other jurisdictions who may not be aware of local innovations.

All this aside, perhaps the biggest limitation of Peer to Patent was that it will only apply to the examination of up-and-coming patents. This means that the current crop of bad patents will dominate the software development landscape for the next 20 years.

As of August 2017, it is unclear whether this program is continuing. Although the program still has a website, the most recent information this author was able find about the project (amongst various broken links), was a second pilot program at the USPTO, that was to end in September 2011.[47]

Ultimately, it is submitted that improved examination procedures, whilst always to be welcomed, amount to little more in this context than re-arranging the deck chairs on the *Titanic*. As with the disclosure reforms, such changes don't answer the criticism that this is not "inventive" activity, in the sense of being within the useful arts.

Compulsory licensing: Compulsory licensing is one means of addressing the problems which the power to exclude causes in an industry characterised by sequential innovation and a high component-to-product ratio. Theoretically, the availability of a compulsory licence limits the ability of a patent holder to hijack a competitor's business by refusing to licence a patented invention, by allowing the competitor to obtain a licence from the Crown. However, the need to pay a 'reasonable' licence fee, and a requirement of a court order mean the transaction costs in using such a scheme are impracticably high. Further, compulsory licences do little to improve the quality or scope of awarded software patents.

Exemptions from infringement: Other solutions borrow from the copyright approach, in order to recognise a distinction between unfair imitation and

---

44 Jason Schultz, attorney for the Electronic Frontier Foundation. Quoted in Daniel Terdiman, "Web Could Unclog Patent Backlog," Wired News, 14 July 2005 <http://www.wired.com/news/technology/0,68186-0.html> (23 August 2006).

45 Schultz, quoted in Terdiman above n44.

46 See R. Demsyn, "Wal-marts Wikipedia War," on *Whitedust Security*, 28 April 2006, <http://www.whitedustnet/article/55/Wal-marts_Wikipedia_War/> (23 April 2006); "Wikipedia Criticised By Its Co-Founder," on *slashdot.org*, 3 January 2005, <http://slashdot.org/articles/05/01/03/144207.shtml>; David Mehegan, "Bias, sabotage haunt Wikipedia's free world," *The Boston Globe*, <http://www.boston.com/news/nation/articles/2006/02/12/bias_sabotage_haunt_wikipedias_free_world/>.

47 Peer Review Pilot FY2011 <https://www.uspto.gov/patent/initiatives/peer-review-pilot-fy2011> (6 August 2017).

legitimate reverse engineering. O'Rourke advocates creating a 'fair use' defence, along the lines of that available in US copyright law.[48] According to this reform, courts would have the power to balance innovation incentives against social benefits in the context of a particular case to determine whether an otherwise infringing activity should be allowed. Such a reform could perhaps bring the patent system more into line with cumulative innovation, and to acknowledge the importance of openness to software development. However, there is a fair amount of uncertainty inherent in the operation of this scheme in that the party claiming fair use must be prepared to brave a patent infringement suit in order to establish their entitlement to continue in their activity.

Another possibility is to explicitly limit the interpretation of patent claims so as to exclude independently implemented functional equivalents.[49] This may be another path to the same objective as the stricter disclosure reform outlined above. Thus it shares the likelihood that it will result in the under-protection of software inventions.

Private initiatives: A number of private initiatives also seek to address the patent problem. IBM, in 2005, started the ball rolling by pledging not to enforce 500 of the patents in its portfolio against Free and Open Source Software (FOSS) developers, and promised to enforce their patents against any organisation pursuing infringement action against an OSS project. Similar pledges have since been made by Sun, Nokia, Novell, Red Hat, and Computer Associates.[50] Putting aside questions of whether these pledges are just publicity stunts,[51] such moves are to be applauded, in that they provide open source developers with at least some level of protection against infringement suits. The problem with such initiatives is that they cannot prevent so-called 'patent trolls' from enforcing their patents against FOSS developers. The typical defensive patent portfolio strategy, which such schemes extend to the FOSS community, is ineffective against trolls, since they do not generally engage in software development and are hence safe from cross-claims for infringement.

Another initiative seeks to raise standards in future software patents by addressing the prior art gap. The Open Source as Prior Art project[52] hopes to improve patent quality by "improving accessibility by patent examiners and others to electronically published source code and its related documentation as

---

48 Maureen A O'Rourke, 'Toward a Doctrine of Fair Use in Patent Law' (2000) 100 *Columbia Law Review* 1177.

49 Julie E Cohen and Mark A Lemley, 'Patent Scope and Innovation in the Software Industry' (2001) 89 *California Law Review* 1 <http://www.law.georgetown.edu/faculty/jec/softwarepatentscope.pdf> (2 November 2004).

50 Robin Cover (ed), "Open Source Development Labs (OSDL) Announces Patent Commons Project" *Cover Pages*, 10 August 2005, <http://xml.coverpages.org/ni2005-08-10-a.html> (13 September 2011).

51 The motives of IBM in particular should be questioned, given the small size of their pledge compared to the size of their portfolio, and the prominence of their role in pushing the pro-IP big business agenda.

52 <http://www.linuxfoundation.org/programs/legal/osapa> (27 August 2011).

a source of prior art."[53] This is a commendable effort to address one of the root causes of poor patent quality in software, yet it is far from a complete solution, and it can do nothing to resolve the problems with dubious patents which have already been awarded.

A final strategy directs the global collaboration model of FOSS development at defeating bad patents. The Public Patent Foundation[54] is a not-for-profit organisation which volunteers to search for prior art which will invalidate bad patents. The Foundation had some successes in challenging patents on the JPEG image format,[55] the drug Lipitor[56] and the FAT filesystem.[57] However, this is a mere drop in the vast ocean of software patents, and amounts to little more than "swatting mosquitoes to cure malaria".[58] As with the Peer to Patent project, although the Public Patent Foundation maintains a website, it appears the project may have lost its steam. The last published article on that site dates from 2015.[59]

Summary: With the exception of the examination improvements, the reforms discussed so far also face a deeper problem. They are technology-specific solutions aimed at tweaking the patent system to meet the needs of the software industry. On a practical level, the consequence of this is that these reforms are likely to face strong opposition from the powerful lobby groups of other industries where the patent system appears to work well, most notably the pharmaceutical industry.

B   Giving effect to the purpose of patent law

Patentable subject matter is also a necessary part of patent law because of the way it gives effect to the purpose of patent law. This purpose is to "benefit society through optimising innovation and public access to new technologies."[60] As the strongest form of monopoly rights, patents can be either significant incentives to innovate, or significant barriers to competition in an industry. As an exception to a general distaste for monopolies, patents should only be available where such a purpose can be served.

Patents are sometimes justified on purely economic grounds. Patent theory asserts that "where the norm of free competition would result in free riding by

53 "Overview," <http://osapa.org/information.html> (4 June 2006).
54 See <http://pubpat.org/> (18 January 2005).
55 See <http://pubpat.org/Chen672Rejected.htm> (1 June 2006).
56 See <http://pubpat.org/LipitorPatentNarrowed.htm> (1 June 2006).
57 See <http://pubpat.org/Microsoft_517_Rejected.htm> (1 June 2006).
58 Richard Stallman, 'The Danger of Software Patents' 2004 Cyberspace Law and Policy Seminar (audio recording), Sydney, 14 October 2004 <http://images.indymedia.org/imc/sydney/stallman%20lo.ogg> (11 November 2004).
59 <http://pubpat.org/> (6 August 2017). The content is a link to this article: Todd Moore, "Why Congress must ensure 'game over' for patent trolls" <http://thehill.com/blogs/congress-blog/technology/236346-why-congress-must-ensure-game-over-for-patent-trolls> (6 August 2017).
60 Advisory Council on Intellectual Property, *Patentable Subject Matter* (Issues Paper, July 2008) ("ACIP 2008") at 1.

competitors and less reason to invest in new technologies, innovation is encouraged by providing innovators with the exclusive rights to their inventions."[61] Innovation is considered to be a positive on the basis that the public benefit from the disclosure of new and useful technologies, which fall into the public domain at the expiry of the patent grant period – the social contract theory of patent law.[62] These benefits must be contrasted with the award of monopoly rights, which amount to a tax on society

> in two different ways: first, by the high price of goods, which, in the case of a free trade, they could buy much cheaper; and, secondly, by their total exclusion from a branch of business which it might be both convenient and profitable for many of them to carry on.[63]

A narrow view of this social contract theory frames the patentable subject matter inquiry in purely economic terms. Yet there is nothing in social contract theory which requires that the benefit received by the public is be limited to economic, or even strictly utilitarian matters. It is submitted that in fact it is impossible to exclude these considerations from patent law, and attempting to do so only converts relevant policy considerations into inherent assumptions.[64] The role of policy in patent law is clearest in Europe, where the various categories of exclusion were considered by the framers to contain their own important policy issues.[65] This point was noted by Laddie J in

---

61 ACIP 2008 at 1.

62 This line of thinking emerged *Liardet v Johnson* (1778) 1 Carp Pat Cas 35 (NP), discussed in Chapter 2. See also *Turner v Winter* (1787) 19 Eng Rep 1276 where the Court noted that "[t]he consideration, which the patentee gives for his monopoly, is the benefit which the public are to derive from his invention after his patent is expired". However, economic rationales for patent law began to dominate as the eighteenth century progressed. See Edward C. Walterscheid "The Early Evolution of the United States Patent Law: Antecedents (Part 4)" (1996) 78 Journal of the Patent and Trademark Office Society 77 at 104– 106. Similarly see the second reading speech for the *Patents Act 1990* Australia: "[t]he essence of the patent system is to encourage entrepreneurs to develop and commercialise new technology.": Commonwealth of Australia, "Patents Bill 1990: Second Reading" Senate, 29 May 1990 <http://parlinfoweb.aph.gov.au/PIWeb/view_document.aspx? id=562046&table=HANSARDS> (2 November 2004). A different theory underlies the US system, namely patents as a reward of inventors: Edward C. Walterscheid "Patents and Manufacturing in the Early Republic" (1998) 80 *Journal of the Patent and Trademark Office Society* 855 at 856.

63 Adam Smith, *An Inquiry into the Nature and Causes of the Wealth of Nations* (5th ed, Methuen, 1904) at 159–160.

64 "[R]ather than asking whether and how patent rights actually encourage inventive activity and dissemination of innovations, courts have simply presumed that they do so in the absence of, and sometimes against, actual evidence.": Richard Gold, "The Reach of Patent Law and Institutional Competence," (2003) 1 *University of Ottawa Law and Technology Journal* 263 at 277. This is what Gold calls "stealth libertarianism". On this approach, difficult questions of policy are reduced to mere technical questions of statutory interpretation. The *Bilski* majority is a case in point.

65 Pila, above n17 at 185.

Fujitsu,[66] although his Honour's approach was ultimately rejected on appeal.[67] Pila advocates for the adoption of such an acknowledgment,[68] noting that it is "consistent with the increasing support for tribunals' reliance on external rules, including fundamental rights, in an effort to stem "the continuing expansion of intellectual property rights outside their traditional bounds" and thereby "correct the slide towards protection [and] reestablish the proper balance of interest."[69] Similarly, Burk and Lemley note the influence of technology specific policy considerations in the development of patent law in the US.[70]

Whether considered on a broad or narrow view, one purpose of patent law is to promote innovation. Whether patent grants are appropriate in a particular field depends on whether that purpose can be achieved. The subject matter test stands apart from the technically-focused tests of inventive step, novelty and utility, in that it alone is directed to the appropriateness of assuming the patent grant will enhance innovation in the field.

In this context, the three-dimensional framework has the advantage that it encourages a direct examination of the nature of the creative process that gives rise to the claims, to assess whether that process is of the kind which is likely to be enhanced by the award of patent rights. As such, it is an analysis which increases the likelihood that the purposes of patent law will be achieved.

The framework has a further important aspect. Despite the "excitingly unpredictable"[71] nature of patent law, the need for flexibility, and however much "times change"[72] it is suggested that this conservative approach, centred around a narrow understanding of the scope of technology, is a proper way to consider the subject matter issue.[73] Whilst it is important that patent law be flexible enough to meet new circumstances, it should not be assumed that patent law

66 "The types of subject-matter referred to in section 1(2) are excluded from patentability as a matter of policy. This is so whether the matter is technical or not.": *Fujitsu Ltd's Application* [1997] RPC 608 (*"Fujitsu"*) at 614.
67 *Fujitsu* per Aldous LJ.
68 Pila, above n17 at 183–187.
69 Pila, above n17 at 185, citing Christophe Geiger, "'Constitutionalising' Intellectual Property Law? The Influence of Fundamental Rights on Intellectual Property in the European Union" (2006) 37 *International Review of Intellectual Property and Competition Law* 371.
70 For example, Burk and Lemley note "an increasing divergence between the rules themselves and the application of the rules to different industries": Dan L. Burk and Mark A. Lemley, "Is Patent Law Technology Specific?" (2002) 17 *Berkeley Tech. Law Journal* 1155 at 1158, and that and that the acknowledgement of that fact requires "courts to build industry-sensitive policy analysis into their decisions": Dan L. Burk and Mark A. Lemley, "Policy Levers in Patent Law" (2003) 89 (7) *Virginia Law Review* 1575 at 1630. Cf. R. Polk Wagner, "Of Patents and Path Dependency: A Comment on Burk and Lemley" (2003) 18 *Berkeley Technology Law Journal* 1341.
71 *NRDC* at 271.
72 *Bilski* at 3227 per Kennedy J (writing for a plurality).
73 Cf. *Bilski* at 3228 per Kennedy J, noting that "new technologies may call for new enquiries", and proposing to modify the patent regime to accommodate the new Information Age.

is "like a nose of wax which may be turned and twisted in any direction",[74] without consequence. As the history of patentable subject matter discloses, and the analysis of the physicality of mathematics and software in Chapters 4 and 5 respectively show, the physicality requirement is both a link to the traditional foundation of patent law, and a necessary limiting device in contemporary understandings of patentability. Similarly, the distinction between the useful and fine arts is a long standing one, which ought be maintained. If the non-physical, expressive and aesthetic arts of the Information Age are to be protected, it is not at all clear, nor should it even be presumed, that patent law provides the best mechanism to do so. It may well be that the new paradigm deserves a new regime to protect it which is contoured to the new balance between freedom and control which such arts require to encourage advancement.[75] Warping an industrial age mechanism designed for industrial era invention is not the solution.

In light of this globally conservative approach, it is also suggested that conservatism at the micro-level is appropriate. Working from a conservative standpoint here means that new areas should be presumed not to be patentable. Such an approach has a number of advantages. It puts the burden of proof of the need for patent protection on the well-organised, well-represented business organisations who are the primary customers of patent offices, instead of the unorganised and largely unrepresented public. It would recognise the cumulative nature of innovation in the knowledge economy, and signal a return to the notion that the "the noblest of human productions – knowledge, truths ascertained, conceptions, and ideas – become, after voluntary communication to others, free as the air to common use."[76] It would encourage public consultation and empirical evidence gathering about the impact of patenting in new areas on the public benefits and detriments *before* such patents began to be awarded. As Gold notes, for courts to do otherwise is to assume competence over difficult questions of policy that they are ill-equipped to address.[77] Should a conservative interpretation of the scope of patent law create a gap which would-be patent applicants suggest needs to be filled, it is appropriate that either Parliament, or an administrative authority such as that considered above, undertake the relevant inquiry into the appropriateness of allowing patents in a new area.

---

74 *White v Dunbar* 119 US 47 (1886) at 51.
75 The merits of a *sui generis* regime for software are considered below.
76 *International News Serv. v Associated Press*, 248 U.S. 215, 250 (1918) (Brandeis, J, dissenting). On a contemporary defence of this position, see Yochai Benkler, "Free as the Air to Common Use: First Amendment Constraints on Enclosure of the Public Domain," (1999) 74 *New York University Law Review* 354 <http://www.yale.edu/lawweb/jbalkin/telecom/benklerfreeastheairtocommonuse.pdf> (15 December 2006); James Boyle, "The Second Enclosure Movement and the Construction of the Public Domain" (2003) 66 *Law and Contemporary Problems* 33.
77 "[G]iven the multiple and multifarious competing interests at stake in determinations of patent eligibility, the judiciary lacks both the capacity and the competence to resolve such issues": Gold, above n64 at 283.

4.2   Characterisation is key

The main weakness of the three dimensional framework is one that it shares with any subject matter test – potential mischaracterisation. The effectiveness of any analysis of an invention will depend on an accurate assessment of what the applicant has actually invented. Proper characterisation, it is submitted, depends on looking behind the form of the claims to their substance,[78] to answer the question from *Grams*, "what did [the] applicants invent?"[79]

This seems to be handled both best and worst under European Patent Convention. The worst approach, being one which exalts form over substance, is that of the European Patent Office, who in allowing claims involving "any hardware" seem to take the word of the applicant as to what they have invented. The best approach is in the UK, where the list of excluded matter, although problematic, forces the recognition that "[i]t cannot be permissible to patent an [excluded item] under the guise of an article which contains that item."[80] To the extent that it is necessary to set out a proper approach to subject matter, a modification of the *Aerotel/Macrossan* case[81] is a good starting point:

1   properly construe the claims;
2   identify the actual contribution;
3   ask whether it falls solely within the excluded subject matter; and
4   check whether the actual or alleged contribution is actually technical in nature.[82]

It is in the fourth step that the three dimensional framework comes into play, the question of whether an alleged contribution is technical in nature being the same as asking whether it falls within the fine arts, or is a technology. It will only be in the EU that the third step is necessary, to take account of the express exceptions in the EPC.

4.3   Generalisations *and* case-by-case analysis are required

Generally, speaking, there will be times where the desirability of awarding patents can be assessed for a whole field, but there will inevitably be exceptions in the form of boundary issues. The case in point in this respect is software. Although it may be clear what the process of developing software is, the growing ubiquity of computing has seen software components become commonplace in all manner

---

78 *Fujitsu* at 530–531 per Laddie J.
79 *In re Grams* 888 F.2d 839 (1989) at 839. See also *CyberSource* at 17: "Regardless of what statutory category ("process, machine, manufacture, or composition of matter," 35 U.S.C. [notdef] 101) a claim's language is crafted to literally invoke, we look to the underlying invention for patent-eligibility purposes."
80 *Merrill Lynch's Application* [1989] RPC 561 at 569.
81 *Aerotel* at [40].
82 *Aerotel* at [40].

of devices.[83] Thus it may be possible for a particular innovation to be characterised in multiple ways, meaning that it falls "within" a number of subject matters. The European experience seems to confirm that even the most explicit blanket bans on patents in particular fields are not guaranteed to work, because of the ease with which alternative characterisations, creative drafting, and generous interpretations of the statutory language work around them.

This is further exacerbated by the fact that the development of new fields of technology often arises in the context of existing areas of study.[84] In this situation, it may not be practical to assess the patentability of an emerging field of science until it becomes 'recognised', by which time the award of broad patents over basic research in the field may have already done significant damage.[85] So to some extent, it will almost always be necessary for courts to determine patentability on a case-by-case basis. Even within the analysis of software in Chapter 5, it was conceded that there might be occasions where claims which might be broadly categorised as claims to software, could be patentable.

However, generalisations about particular types of inventions are instructive in two regards. First, they allow a degree of certainty for innovators in that they can make some sort of assessment of the likelihood of patent protection being available before entering into the expensive and time consuming process of applying for a patent. A degree of certainty also benefits those tasked with examining the patent for validity. Second, they reflect the nature of judicial development of the law – courts determining the patentability of a claimed invention will always look for guidance from past cases, and will be influenced by the patentability of similar, or even analogical inventions. Since Australia is not a heavily litigated jurisdiction, this can lead to long periods in which the current state of the law is out of step with the technology it is supposed to regulate. For example, although the patentability of software first came to the attention of the US Supreme Court in 1972 in *Gottschalk v Benson*,[86] it was 1991 before the patentability of software was judicially determined in Australia in *IBM v Commissioner of Patents*.[87] In the meantime, it was left to the Australian Patent Office to provide guidance to patentees by updating its Manual of Practice and Procedures to reflect the development of the law in other jurisdictions.[88]

---

83 For example, is an innovative method of curing rubber, such as that in *Diamond v Diehr* which uses software to determine the heat of the mould an innovation in the field of software development, or in materials engineering?

84 The emergence of computer science from the field of mathematics is just one example.

85 See for example, in relation to biomedical research, Michael A. Heller and Rebecca S. Eisenberg, "Can Patents Deter Innovation? The Anticommons in Biomedical Research" (1998) 280 *Science* 698 at 700–701. Cf. Dianne Nicol and Jane Nielsen, "Patents and Medical Biotechnology: An Empirical Analysis of Issues Facing the Australian Industry" (Centre for Law and Genetics Occasional Paper No 6, 2003) at 89.

86 409 US 63 (1973).

87 (1991) 33 FCR 218.

88 By August 2011 the Australian Patent Office had "updated" the Australian law in light of the developments in *Bilski*, in *Invention Pathways Pty Ltd* [2010] APO 10 (21 July 2010); and the Federal Circuit case of *CyberSource Corporation v Retail Decisions Inc* Appeal No 2009-1358

It may well be that the appropriate middle ground between case-by-case analysis and per-field determinations of technology lies in some form of administrative solution.[89] In this respect, the development of authoritative guidelines, rather than a pure legislative solution, is the appropriate way of moving forward. In addition to the appointment of bodies to undertake *ad hoc* reviews of an area,[90] expert advisory committees could (and should) be available to assist with the examination of difficult patents, the development of guidelines, and to conduct comprehensive reviews of areas of particular concern on an ongoing basis. Such bodies might properly consider the full range of economic, social, and ethical considerations relevant to the subject matter issue to be resolved, engaging in public consultation as a part of the process. The decisions of this body may be subject to ordinary judicial review, and may also be given some statutory weight in judicial determinations of subject matter issues.

4.4 Human rights jurisprudence

There is a growing awareness of the tension between intellectual property rights and human rights such as rights to adequate health care, to education, to share in the benefits of scientific progress, and to participation in cultural life.[91] Such considerations are viewed by some as external to the concerns of the patent regime, and somehow irrelevant to the development of patentable subject matter

---

(Fed Cir, 2011) in *Network Solutions, LLC* [2011] APO 65 (19 August 2011). For a critical summary of these developments, see Mark Summerfield, "Australian Patent Office Shoots Down Another 'Business Method'" on *Patentology* <http://blog.patentology.com.au/2011/08/australian-patent-office-shoots-down.html> (28 August 2011). Similarly, it might be argued that the Full Federal Court decisions in *Research Affiliates v Commissioner of Patents* (2014) 227 FCR 378 and *Commissioner of Patents v RPL Central Pty Ltd* [2015] FCAFC 177 "updated" Australian law to be consistent with more recent developments in *Alice Corp v CLS Bank International* 134 S. Ct. 2347 (2014).

89 William W. Fisher III, *Promises to Keep: Technology, Law and the Future of Entertainment* (Stanford University Press, 2004) at 195.

90 Despite the US Presidential Commission which recommended against software patenting in 1966 being made up of "distinguished scientists, academics, and representatives of leading computer and high technology firms (as well as the Commissioner of Patents)": Pamela Samuelson, "Benson Revisited: The Case Against Patent Protection for Algorithms and Other Computer Program-Related Inventions," (1990) 39 *Emory Law Journal* 1025 at 1038, the recommendations were never acted upon.

91 See for example Laurence R. Helfer, "Human Rights and Intellectual Property: Conflict or Coexistence?" (2003) 5 *Minnesota Journal of Law, Science and Technology* 47; Gold, above n64 at 283–284, noting the need to consider the social, ethical and economic implications of technology in order to guarantee its just use; James Boyle, "Enclosing the Genome: What the Squabbles over Genetic Patents Could Teach Us" in F. Scott Kieff, *Perspectives on the Human Genome Project* (Academic Press, 2003) 97, recognising that the patenting of genes involves a range of considerations beyond a narrow utilitarian inquiry. The impact of patent grants on public health is recognised by the *Doha Declaration on TRIPS and Public Health*, WTO Doc WT/MIN (01)/DEC/2 (2001) which allows Members to flexibly interpret the patent obligations in TRIPS to allow for access to medicines.

principles.[92] However, it is submitted that many of these rights are fundamental to a proper understanding of the traditional limits on patentable subject matter. In Chapter 4, it was shown how the notions of freedom of thought and freedom of expression, fundamental rights so familiar to the human rights lawyer, inform an understanding of why mathematical advances are in the fine rather than the useful arts.

One possible approach in this book would have been to frame an argument against patentability based in human rights.[93] That course was not taken. Nonetheless, it is clear that the framework bears more than a passing relationship to these freedoms. That this is so is obvious in the distinction drawn between expression and purpose. Where the fine arts are involved, the importance to be placed on freedom of expression is greater, as the expression itself is a source of value. Where an activity is purposive, expression must submit to function, and therefore the need for freedom of expression is lessened. In other words, the need for freedom in a particular discipline may aid in determining whether a particular art is better categorised as a fine art or a useful art. A similar point can be made in relation to the intangible/physical distinction. Where abstract concepts are involved, and the physical realisation of such concepts is only of secondary significance, freedom of thought is of fundamental importance. As a corollary, if an activity is directed by, or heavily reliant upon human thought, then it is more likely that a fine art is involved.

The aesthetic/rational dimension is more difficult to neatly characterise. Aesthetic responses are described as emotional, intuitive, and highly subjective. This of itself suggests the importance of the individual, and in particular the individual mind. This of course invokes the need for freedom of thought. But as discussed in Chapters 4 and 5, such reactions are both provoked by expression, and aesthetics also motivates the creation of expressive works. It may be that the aesthetic operates at the intersection of expression and thought. As such the importance of thought and expression together will tend to support a classification of subject matter as falling within the fine arts.

Understanding that the freedom operates as a consideration in the patentable subject matter dispute naturally opens the door to a consideration of the intersection of intellectual property and human rights jurisprudence. It is acknowledged that "the historical connections between intellectual property and human rights are thin at best."[94] As noted above, justifications of the patent regime are often couched in economic or utilitarian terms. As a result, "patent law has been almost entirely isolated from First Amendment considerations".[95] Yet to deny the applicability of human rights in the patent context is wrong,

---

92 See for example *Grant v Commissioner of Patents* (2006) 154 FCR 62 at 72. See also Boyle, above n91 at 106–109.
93 At an early stage of the PhD thesis which gave rise to this book, just such an argument was envisaged.
94 Peter Drahos "Intellectual Property and Human Rights" [1999](3) *Intellectual Property Quarterly* 349 at 357.
95 Dan L. Burk, "Patenting Speech" (2001) 79 Tex. L. Rev. 99, at 137.

because it ignores one of the fundamentals of property law – that awarding property rights does not just give the holder a power to control the thing which is the subject of the grant, it also gives the holder a power over other people.[96]

In Chapter 4,[97] it was noted in passing that a human rights characterisation of patent law is considered by some to be paradoxical; as an irreconcilable struggle "when one human right is pitted against another, when intellectual property rights are used to restrict access to information that could – at no real cost to the developer – be deployed in ways that satisfy fundamental needs".[98] However, rather than characterising the intersection as a battle between the conflicting spheres of human rights and intellectual property, it is submitted that it is of more benefit to look within the patent paradigm to determine how human rights values capture and explain the "multiple and multifarious competing interests at stake in determinations of patent eligibility".[99]

The trend towards a narrow inquiry, including an unwillingness to consider the effect of patent grants on human rights, is not evident in copyright law. In fact, copyright law has been claimed to be "the engine of free expression".[100] The history of copyright in the US jurisdiction reveals an understanding of the links between copyright and the pursuit of democracy.[101] In 1790, the Senate committee noted that "literature and science are essential to the preservation of a free Constitution"[102] namely because of the way it supports the diffusion of knowledge amongst the electorate.

In addition, freedom of thought arguably finds its place at the boundaries, through the idea/expression dichotomy,[103] and the related doctrines of

---

96 See for example Morris R. Cohen, "Property and Sovereignty" (1927) 13 *Cornell Law Quarterly* 8 at 13, who characterises property as a sovereign power: "[W]e must not overlook the actual fact that dominion over things is also *imperium* over our fellow human beings."

97 Section 2 on page 125.

98 Rochelle C. Dreyfuss, "Patents and Human Rights: Where is the Paradox" in William Grosheide (ed), *Intellectual Property and Human Rights: A Paradox* (Edward Elgar, 2010) 72 at 72.

99 Gold, above n64 at 283. See also Dreyfuss, above n98 at 73.

100 *Harper & Row, Publishers, Inc. v Nation Enters* 471 U.S. 539 (1985) at 558.

101 As noted in Chapter 4, one of the justifications of freedom of expression is that it promotes democracy. See Chapter 4, at 138 and following.

102 Bruce W. Bugbee, *Genesis of American Patent and Copyright Law* (1967) at 137, cited in Neil Weinstock Netanel, "Copyright and a Democratic Civil Society" (1996) 106 *Yale Law Journal* 283 at 289.

103 See *Baker v Selden* 101 U.S. 99 (1879) (*"Baker v Selden"*) at 100–01: "[T]he truths of a science or the *methods of an art* are the common property of the whole world, and [the] author has the right to express the one, or explain and use the other, in his own way." (emphasis added); *Hollinrake v Trusswell* [1894] Ch 420 at 427 per Lindley LJ: "Copyright ... does not extend to ideas, or schemes, or systems, or methods; it is confined to their expression; and if their expression is not copied the copyright is not infringed." In *Autodesk v Dyason (No.2)* (1993) 176 CLR 300, Mason J described the idea/expression divide as "the dominant principle in copyright law". Dawson J in the same case noted that "when the expression of any idea is inseparable from its function, it forms part of the idea and is not entitled to the protection of copyright. See also *IceTV v Nine Network* (2009) 239 CLR 458 at [22]–[28] per French CJ, Crennan and Kiefel JJ.

merger[104] and *scènes à faire*.[105] Both of the latter doctrines are of particular relevance to determining the copyrightability of software in the US, being key components of the second stage of the *Altai* abstraction-filtration-comparison test.[106]

Even in Australia, with its limited conception of freedom of speech,[107] the idea versus expression dichotomy has been directly linked with freedom of expression. In *Skybase Nominees*, Hill J in particular noted the

> tension in policy between the monopoly rights which are conferred upon the owner of copyright in a literary, dramatic or artistic work on the one hand, and the freedom to express ideas or discuss facts on the other. While there will be an infringement of the copyright of an owner in a literary, dramatic, musical or artistic work where there is a reproduction of that work or a substantial part of it, the fact that another work deals with the same ideas or discusses matters of fact also raised in the work in respect of which copyright is said to subsist will not, of itself, constitute an infringement. *Were it otherwise, the copyright laws would be an impediment to free speech, rather than an encouragement of original expression.*[108]

What is suggested in relation to the patent paradigm then is that there is plenty of room for human rights theory within the traditional bounds of the patent law paradigm. The fact that a framework such as the one developed, which is

---

104 The doctrine of merger holds that where only a limited number of ways of expressing an idea exist, the expression merges with the idea and becomes unprotectable. See *Baker v Selden* at 103: "Where the art [that the book] teaches cannot be used without employing the methods and diagrams used to illustrate the book, or such as are similar to them, such methods and diagrams are to be considered as necessary incidents to the art, and given therewith to the public."

105 The doctrine of *scènes à faire* operates to remove from copyright "those elements which are standard and inevitably arise in the treatment of a given topic": Stanley Lai, *The Copyright Protection of Computer Software in the United Kingdom* (Hart Publishing 2000) 41 at [3.1].

106 See *Computer Associates Intl, Inc v Altai Inc* 982 *F.2d* 693 (1992) at 709–710. The court noted that the doctrine of merger is particularly relevant in the context of software, since:

> a programmer's freedom of design choice is often circumscribed by extrinsic considerations such as (1) the mechanical specifications of the computer on which a particular program is intended to run; (2) compatibility requirements of other programs with which a program is designed to operate in conjunction; (3) computer manufacturers' design standards; (4) demands of the industry being serviced; and (5) widely accepted programming practices within the computer industry; such "external factors" may make expression unprotectable.

107 Specifically, all that is recognised is an implied freedom of political communication, which acts not as a positive right but a limitation on legislative power. See *Lange v Australian Broadcasting Corporation* (1997) 189 CLR 520.

108 *Skybase Nominees Pty Ltd v Fortuity Pty Ltd* (1996) 36 IPR 529 at 531 per Hill J (French J concurring) (emphasis added).

consistent with the theory of patent law, and which opens the door to such considerations can be developed, suggests that other evaluative tools might be similarly developed to address other ethical and social values which patent grants inevitably affect.

5   Beyond the patent paradigm

If software is to be pushed outside the patent regime, the question arises as to how software ought to be protected, if it is to be protected at all. Whilst an answer to that question could well provide enough material for another thesis, it is apposite to say something about how the software landscape might look in the absence of patent protection.

It is often assumed that inventors will have no way of stopping competitors from imitating their innovations. Although it is impossible to measure how many competitors would copy an unprotected innovation in the absence of patent protection, some indication of the likely extent of the problem can perhaps be gleaned by looking at the extent to which copying can be found in patent infringement cases. Despite proof of copying not being a requirement of infringement, it seems fair to conclude that those who would copy an invention *with* patent protection form a subset of those who would copy without patent protection. It might then be expected that a significant proportion of infringement actions would involve copying rather than independent invention. However, a 2008 study by Cotropia and Lemley[109] found allegations of copying (or facts that suggested that copying may be in issue) in 11% of cases,[110] and copying was actually found in only 1% of cases reviewed. This factor varied by industry, with the high water mark being the pharmaceutical industry, where allegations of copying were found in 65% of cases (no doubt due to the existence of generic producers). Computer-related inventions and software on the other hand, despite forming the largest subset of the cases considered, involved an extremely low rate of allegations of copying in only 2.6% and 3% of cases, respectively.[111] So it seems that the free rider problem may be less of an issue than it is usually given credit for, at least outside the pharmaceutical industry, where the importance of patent protection is generally uncontested. These figures certainly indicate, if nothing else, that presumptions about the scale of the free rider problem cannot be generalised.

The size of the free riding problem will also depend upon the extent to which alternative protection regimes are available. Alternative protection is typically assumed to mean trade secret protection, although it may often be the case that copyright could also be available. This is certainly true in the software context, where those against software patenting often claim that copyright

---

109 Christopher Cotropia and Mark Lemley, "Copying in Patent Cases (draft)" <http://www.law.berkeley.edu/institutes/bclt/students/2008_ip-seminar/Lemley_Copying-in-Patent-Law1.pdf> (17 September 2008).
110 Cotropia and Lemley, above n109 at 24.
111 Cotropia and Lemley, above n109 at 27.

provides adequate protection.[112] The availability of copyright as an alternative protection paradigm may explain the low rate of copying in patent infringement suits noted above. On this note, the possibility of establishing a software-specific protection regime is considered below.

Software itself falls to be protected by more than just the patent regime. There is a significant overlap between patent, copyright, and trade secret protection. All three paradigms are used to varying extents by software developers, and software companies, to protect their creations. The nature of the protection afforded by those other regimes, and the shortcomings of these protections is considered below.

5.1   Trade secrets

Trade secrets are sometimes treated as the antithesis of patent law. Patent law is said to encourage disclosure, where as trade secret law encourages secrecy. By distributing only object code, it is possible for the informational aspects of software to be obscured from view. So-called proprietary software depends on a combination of secrecy, and copyright to prevent literal copying. Various obfuscation techniques can also be used to make reverse engineering of software more difficult,[113] as can various legal mechanisms, such as non-disclosure agreements, and anti-reverse engineering clauses in software licences.[114]

The time it takes software to be reverse engineered by competitors provides a definite first mover advantage to innovators, who can in theory move on to the second iteration of a product. That lead time can itself act as an incentive for the first-mover to create further refinements, which mean that their competitors are always one step behind. Further, the very act of reverse engineering is seen to be "an essential part of innovation",[115] as it invites follow-on innovators into the field, which "could lead to significant advances in technology".[116]

---

112 "Software developers are perfectly protected without patents. Everyone who writes a computer program automatically owns the copyright in it. It's copyright law that made Microsoft, Oracle, SAP and the entire software industry so very big. It's the same legal concept that also protects books, music, movies, paintings, even architecture.": *NoSoftwarePatents. com*, "The Basics" <http://www.nosoftwarepatents.com/en/m/basics/index.html> (4 August 2011).

113 For a selection of the sorts of methods used in software, "Code Obfuscation Literature Survey" on *Obfuscators.org*, 13 April 2008 <http://www.obfuscators.org/2008/04/code-obfuscation-literature-survey.html> (1 September 2011).

114 Pamela Samuelson and Suzanne Scotchmer, "The Law and Economics of Reverse Engineering" (2002) 111 *Yale Law Journal* 1575 at 1660. See also Electronic Frontier Foundation, "Coders' Rights Project Reverse Engineering FAQ" <https://www.eff.org/issues/coders/reverse-engineering-faq> (1 September 2011), noting that anti-reverse engineering clauses for software may be found in "End User License Agreements (EULA), terms of service notices (TOS), terms of use notices (TOU), a non-disclosure agreement (NDA)".

115 See *Bonito Boats, Inc v Thunder Craft Boats, Inc* 489 U.S. 141 (1989) ("*Bonito Boats*") at 160.

116 See *Bonito Boats* at 160.

It is sometime claimed that the lead time has shortened to such a degree that trade secrets have ceased to be an effective way of ensuring an appropriate reward for innovative research and development.[117] But the extent to which a redirection from patent to trade secret should be considered a negative consequence depends on the extent to which patent applications amount to a useful disclosure of the invention. In the software context, patent literature is rarely of use to programmers. that the main reason for this is that the abstract nature of patent claims means that they fall on the wrong side of the semantic gap. So it must be questioned whether the absence of patent protection for software-related inventions should be considered a loss at all.

Further the impact of reverse engineering must be considered. It may be that the lead time associated with reverse engineering is more attuned to the market cycle for products than the static 20-year time frame associated with patents. If it is, then the likelihood of disruption of next-generation inventions is not possible in the same way that submarine patents can disrupt an industry by providing protection for other products not within the contemplation of the original inventor, suggesting an overall economic gain may be achieved by such a redirection. Also, reverse engineering of a product may in fact lead to new information about the innovation being discovered which can lead to valuable improvements. Finally, the economic value of patent protection depends on the extent to which independent invention should be viewed as a valuable exercise in its own right – in some fields like software it may be no less than critical.

Regardless of the availability of patents for software, trade secret protection remains an important protective mechanism. It has advantages over both patent and copyright in that it is immediate, does not require registration, is potentially broader in scope, and unlimited in duration. It also has its own technologically sensitive built-in regulation mechanism – reverse engineering. Its suitability is therefore largely determined by the time required for competitors to reverse engineer then re-implement a particular secret. In fact the trade secret paradigm is certainly a feature of the proprietary software sector, who rely on copyright protection to prevent literal copying, and trade secret protection, via object code.[118]

5.2 Copyright

As is clear from the analysis in the previous chapter, the fine arts are the traditional province of copyright law. As such, it makes sense to consider the extent to which copyright provides appropriate protection. It certainly protects

---

117 Jerome Reichman, "Legal Hybrids Between the Patent and Copyright Paradigms" (1994) 94 *Columbia Law Review* 2432; Samuelson et al., above n39.

118 In fact, it is not unusual to use code obfuscation as a technique to extend lead time. See for example, Wikipedia, "Obfuscated code" <http://en.wikipedia.org/wiki/Obfuscated_ code> (11 September 2011). The camel code and DeCSS code in Chapter 5 are examples of obfuscated code although the purpose was not to increase lead time, but for aesthetics.

against literal copying, and is entirely consistent with the freedoms required in an aesthetic, expressive field such as software development. On any reading, the nature of source code as akin to a literary work is hard to deny. Yet, to enable "full" copyright protection, it has been necessary to modify the definition of a literary work to include object code. This has resulted in somewhat of a distorting of the conceptual consistency of a literary work, as object code is not "intended to afford either information and instruction, or pleasure, in the form of literary instruction" to any person, and is unlike other literary works in that way. However, as noted in the previous chapter, the information which is embedded in such code is still available, although it may require decompilation in order to make it available, and at the very least contains a kernel of expression.

The major limit of copyright which prompted a push for the patenting of software still exists however. This limitation is that copyright paradigm only protects the literal text of software, and "the primary source of value in a program is its behaviour, not its text."[119] Although text and behaviour are directly linked, they are not identical. It is possible for any skilled programmer to "copy the behaviour of a program exactly, without appropriating any of its text".[120] In this situation, the 'imitation' will not infringe the copyright in either source or object code, but will behave in a manner indistinguishable to an end user from the original.

However, there is a significant proportion of global software development which depends only on copyright protection. That is the Free and Open Source software sector.[121] Populated by a significant number of volunteers, but also a number of large commercial software development organisations, the FOSS model is different to the proprietary software camps in that it eschews secrecy in favour of the open disclosure of source code. Rather than seeking increased protection, the FOSS model actually increases the freedom available to developers by granting an open licence to reuse, distribute and modify, with the only real limitation being that such freedoms are to be passed downstream.[122] The significance of the FOSS sector was discussed in Chapter 2.[123]

5.3   Between the cracks?

To the extent that suitable protection is not offered neither patent, trade secret, or copyright, it is submitted that this is not a fault of patent law, but an inevitable consequence of the disparate nature of the regimes which fall under the penumbra of intellectual property. Where there is a consistent theoretical framework informing any intellectual property paradigm, it can maintain that consistency.

---

119 Samuelson et al., above n39 at 2315.
120 Samuelson et al., above n39 at 2315.
121 See the discussion of FOSS software and its significance in Chapter 2 at 64.
122 For an overview of the various alternatives, see Kenneth Wong and Phet Sayo, "Free/Open Source Software: A General Introduction," *International Open Source Network*, 2004 <http://www.iosn.net/foss/foss-general-primer/> (18 September 2010).
123 See Chapter 2, on page 64.

And to an extent, that consistency requires that inconsistent subject matter be excluded.

Copyright, according to the *Baker v Selden* exclusion excludes functional subject matter. So whilst software has caused some theoretical weakening of the consistent paradigm, by watering down the notion of a literary work to cover things not intended to afford literary instruction or enjoyment, functional protection is clearly excluded. Similarly, it has been shown that patent law ought to exclude the subject matter of the fine arts. This may mean that software fails to be afforded anything other than the "thin" protection of copyright law – against literal copying – but is denied any form of functional protection.

It may be suggested that such protection has marginal, if any value, and that copyright is falling through a crack between copyright and patent paradigms. To an extent, this may be true, in that neither paradigm is equipped to deal with subject matter that is both descriptive and functional. Given the independent history of each of the paradigms, that this might happen does not represent a failure of either paradigm. It merely suggests that it may be time to consider adding another paradigm to the intellectual property stable.

5.4   An alternative protection paradigm?

Over 15 years ago, a group of leading computer science and intellectual property scholars turned their efforts to a "normative analysis of the kind of legal protection that would be socially desirable for software and how it might best be accomplished".[124] The authors were critical of the way in which discussions of the software protection problem focused on adapting existing regimes, showing how such an approach leads to recurrent cycles of under- and over-protection.[125] Their conclusion was that a *sui generis* regime was the only way to correctly strike the balance between creating incentives to innovate and upholding the public interest.[126] Their proposed regime starts from the principle of "*market preservation*, that is, constructing just enough machinery to head off the ways in which marketplaces fail".[127] The regime chosen to give effect to that principle is based around the following four precepts:

- Traditional copyright protection for literal code;
- Protection against behavior clones for a market-preserving period;
- Registration of innovation to promote disclosure and dissemination; and
- A menu of off-the-shelf liability principles and standard licenses.[128]

---

124 Samuelson et al., above n39.
125 Samuelson et al., above n39 at 2356–2357.
126 Samuelson et al., above n39 at 2356–2357.
127 Randall Davis et al., "A New View of Intellectual Property and Software" (1996) 39 *Communications of the ACM* 21 at 21. Davis is one of the authors of the longer Manifesto article, and this article provides a useful summary of it.
128 Davis et al., above n127.

Copyright protection is an important starting point in software, since it alone was sufficient for the early successes of the software industry.[129] Copyright protection, as the traditional protection regime for the fine arts, is an obvious starting point for an appropriate protection mechanism, given the result of the analysis conducted in Chapter 7. Because it protects innovators against free-riding imitators, whilst not interfering with independent creation, it promotes the freedom which lies at the core of software development, and upon which the analytical framework is based. Copyright also lies at the heart of the FOSS paradigm discussed above. Such a regime would allow FOSS projects to continue with little disruption.

But the absence of functional protection suggests copyright alone may be insufficient. In particular, the analysis of the nature of software in Chapter 5 noted that software contains functional elements, and that function may become a dominant consideration. The *sui generis* system directly addresses the behavioural protection issue because, like the patent regime, it provides some protection against functional imitation, or "behaviour clones". The system works in a similar fashion to the trade secret paradigm by creating an artificial lead-time for the registrant of a grain-sized innovation, in exchange for disclosure of the details of the innovation. All those wishing to use the innovation during the lead-time must pay a royalty to the innovator, or else wait out the liability period.[130]

These behavioural protections have been argued by Reichman to be equivalent to patent protection in that no one is worse off under this system than they would be if software patents were available.[131] Registrants are protected against functional imitation for a period consonant with the market life of their product. Follow-on innovators contribute to the research and development costs of upstream inventors, without the inconvenience of the large transaction costs of licence negotiation. Developers of independent products can stand on their own in the marketplace, without fear of ambush. In addition, the regime upholds the importance of aesthetics. It encourages disclosure, not through

---

129 See for example Richard Stallman, "For submission to the Patent & Trademark Office," 1994, <http://lpf.ai.mit.edu/Patents/rms-pto.html> (13 September 2006); Bill Gates, "Challenges and Strategy" Microsoft Internal Memo, <http://www.bralyn.net/etext/literature/bill.gates/challenges-strategy.txt> (13 September 2006; D. Brotz (Principal Scientist, Adobe Systems), Public Hearing on Use of the Patent System to Protect Software Related Inventions, Transcript of Proceedings Wednesday, January 26, 1994 San Jose Convention Center, <http://www.gordoni.com/software-patents/statements/adobe.testimony.html> (13 September 2006).

130 Reichman has elsewhere observed that many of the major problems facing the patent system are due to a failure to protect these small grain-sized innovations. See Jerome H. Reichman, "Of Green Tulips and Legal Kudzu: Repackaging Rights in Subpatentable Innovation" (2000) 53 *Vanderbilt Law Review* 1753.

131 See Reichman, above n130, wherein he considers the market life of a hypothetical innovation in the plant-breeding industry, showing how the original innovator is adequately compensated for his contribution to the industry, whilst open access facilitates follow-on innovation.

the patent claims and specification, written as they are for the lawyer and not the programmer, but the disclosure is in a familiar form, in code. That disclosure therefore provides access to material by which programmers can develop their evaluative aesthetic. In addition, it may well trigger the generational aesthetic, inspiring the creation of follow-on innovations.

Samuelson et al. did not elaborate on the exact details of how such a system should work, preferring to focus on the framework of the system with a view to "facilitat[ing] and direct[ing] the political debate."[132] A question thus arises as to how to such a market-oriented system could be configured to meet the needs of the local industry. The first stage in such a process would be the gathering of empirical data as to the life cycle of software innovations. Such a process should be enhanced by a consultative process, gathering the views of all interested parties in the national software market. During the initial iteration, it may be advisable to adopt a conservative position, so that those holding software patents are not disadvantaged. Subsequent iterations could be informed by the reporting on information captured during the registration of innovations, which should be freely accessible. This data should be used to regularly (perhaps annually) assess and tweak system parameters, such as innovation shelf-life to match the actual life of innovations.[133]

Davis et al., in introducing the paradigm in 1995, made the following remarks which are just as relevant today:

> Each time there appears to be a lull in the controversy about legal protection for software, we are quickly jolted by the battle being joined anew. The difficulties won't soon disappear, ... because there is a deep seated problem here: existing intellectual property laws are fundamentally ill-suited to software.[134]

The failure to adopt such a model means that the cycle continues anew. As mentioned in the Introduction, the next instalment of the patent wars has started, the battleground this time being smartphones.[135] The battle in Europe is just about to resume after a five-year hiatus, this time over the unitary patent regime. If the software industry is to have any certainty, surely it is time to correct the mistakes of the past. It seems clear that patent law cannot accommodate software's unique nature, and consequent needs. Therefore it is time to move forward and consider how the landscape without software patents might look. This alternative regime provides a glimpse of that future.

---

132 Samuelson et al., above n39 at 2315.
133 On the operation of a similar regime for the protection of copyright, see William W. Fisher III, *Promises to Keep: Technology, Law and the Future of Entertainment* (Stanford University Press, 2004).
134 Davis et al., above n127 at 21.
135 See Introduction, n27 on page 5 and accompanying text.

Conclusion

The primary object of this book was to examine the patentability of software, and in particular to put forward the reasons why software should not be patentable. The argument proceeded from some observations in Chapter 1 about the history of software, and the way in which it was developed. It was seen from the history of software how successive layers of abstraction away from the physical implementation of the computer formed the foundation for new developments, and also allowed programmers to tackle problems of an ever-increasing increase in complexity. The development of software was also demonstrated to be a journey away from the ultimate in abstract artefacts, the idea, towards a (somewhat) less abstract implementation of that idea in software, travelling through various levels of decreasing abstractness towards that goal. The chapter then explored the assertion that mathematics and software are identical, or isomorphic. It was shown that this assertion was substantiated by its historical origins in the logicist and formalist programmes in mathematics. It was also verified by a correspondence between the practice of mathematics, and the practice of software development.

In Chapter 2 it was shown how the patenting of software was hard to justify, both as a practical matter for those developing software, and also as a matter of theory. The way in which courts and tribunals in every jurisdiction have been confounded in their attempts to put forward a compelling explanation of why it is that software should or should not be patentable led to the conclusion that a new approach is required.

The development of just such a new approach began in Chapter 3, where the nature of mathematics, as understood by both mathematicians and patent lawyers was considered. One of the important lessons from the history of mathematics, and attempts to set out a theoretical foundation for the discipline, is that ontological accounts, at least in mathematics, are somewhat of a red herring. After countless centuries, the nature of mathematics has not been resolved. All accounts considered capture something of the essence of mathematics, but all have their problems.

An attempt was made nonetheless to reconcile the status of mathematics within patent law by reference to the traditional exclusions developed. This was done by taking a holistic view, in that an explanation of mathematics' non-patentability should only be accepted if it could be reconciled with all theories of mathematics. It was thereby demonstrated that no traditional exclusion gave a complete account of the non-patentability of mathematics.

Chapter 4 took a different approach, asking not what mathematics was, but what it required for its advancement. The answer to that question was much more fruitful, and led to an investigation of the role of freedom in mathematics. From that explanation, and a consideration of the philosophy of technology, a three dimensional framework of analysis was developed.

From the application of that framework to software in Chapter 5, it was demonstrated why programming was not a useful art, a technology, but a fine art. As far as most of us are concerned, it is the symbolic aspects of software which are most important, not the physical. Whilst it is possible that a computable process might form part of a physical device, claims artefacts existing higher up the stack, such as algorithms and abstract data structures, are so abstract that they ought never to be considered patentable. It was shown that despite the fact that implements a function, the expressiveness of software permeates every level of software development, and every aspect of software even when in executable form. Finally, the logical sequence of steps, and the application of software to "serious" projects from science to business, as well as the desire of some to relive the early days of software and transform software into a form of engineering, the software development process does not follow a rational, sequential path like the idealised waterfall model suggests. Software is evaluated by, motivated by, and generated by, aesthetics. As such, programming is better described as a fine art than a useful art, and would, in the vast majority of cases at least, not be patentable subject matter.

Application of the analysis to software also satisffed a further aim of this thesis – to identify commonly misunderstood aspects of the nature of software. In particular, it was explained how software, which might on a cursory analysis appear to be physical, either by reference to its storage on physical media, or its effect of turning a computer into a new machine, is at its core an intangible artefact. Second, a consideration of the role of the aesthetic in evaluation, motivation, and generation of software was shown to belie the rigorous, logical sequence of steps in which it takes its final form. It is perhaps in this aspect in which the asserted identity between mathematics and software finds its strongest manifestation.

In Chapter 6, the consequences of that position were explored. As a starting point, it was shown how the framework developed gave due consideration to the important features of software noted in Chapter 1. It was then noted that each of the jurisdictions surveyed in the thesis gave at least some consideration to the physicality/intangibility dimension of the analysis, but only lesser consideration to the other two aspects.

The final aim stated at the outset, was to develop a mode of analysis which went beyond a narrow, technical analysis of patent law. It is clear that the framework developed satisffes that goal. Its genesis was an acknowledgment of the role of freedom in mathematics, and it therefore demonstrates how fundamental human rights concepts are at work in patent law, not as an external influence requiring recharacterisation, but as concepts which flow through the development of traditional patent law principles. It also demonstrates how novel investigative tools, directed to other relevant considerations, such as social, moral, and ethical considerations might be developed in a similar manner.

More broadly, it was argued that the need to exclude software from patentability, and the distinction between fine arts and useful arts, suggested that the subject matter inquiry has continuing relevance as a gatekeeper in patent law. No amount of tinkering with the other technical tests, examination regime improvements, or private initiatives are likely to be successful. The role of the inherent patentability inquiry is unique within patent law, as the only way in which patent law can properly engage with policy, ethical, and social issues upon which the patent system has an impact. The relationship of the framework to the freedoms of thought and expression demonstrates how it is possible to craft an analytical tool which is consistent with the history of patent law, but also allows room for these "soft" issues to be considered in a structured way.

Finally, the implications of software's non-patentability on its protection were considered. Removing patents as a protection mechanism for software does not leave a gaping hole. There is evidence suggesting the free rider problem is not as large as the pro-software-patent lobby might suggest. A variety of protections beyond patent law are presently used to protect different aspects of software. Either alone or in combination, it is possible to imagine how remaining paradigms provide protection to software which is appropriate to its nature. This is not a theoretical claim, but one borne out of the history of software before the award of software patents.

To address any remaining concerns about the gap in protection, an alternative protection paradigm was discussed. This paradigm, based as it is on a contemporary analysis of factors likely to fulfil the needs of innovators and competitors is one which is much contoured to the needs of this Information Age creation.

A concluding comment

As computer hardware gets smaller, faster, more connected, and more reliable, the software which controls it grows becomes ever more present in our daily lives. New modes of interaction, once the stuff of science fiction, such as voice recognition, motion sensing, face recognition, and increased connectivity change the way we access information, connect with each other, work together, and play together. New advances in software have a central role to play in this context, and as such software undoubtedly has great promise. But it flows from both the reach of software, and its potential benefits, that we must consider carefully how best to encourage its advancement. Patent law is but one potential mechanism by which innovation might be encouraged. It has a long history, but it reflects certain policy choices about a balance between freedom and control which are connected to the types of physical inventions which have been within the historical scope of patent law. As software does not fit within that category, it ought not be patentable, and a new paradigm must be found to properly balance its unique requirements.

Index